Aluminum in America

ALSO BY QUENTIN R. SKRABEC
AND FROM MCFARLAND

*Benevolent Barons: American Worker-Centered
Industrialists, 1850–1910* (2015)

Rubber: An American Industrial History (2014)

*The Green Vision of Henry Ford and George Washington Carver:
Two Collaborators in the Cause of Clean Industry* (2013)

*The Carnegie Boys: The Lieutenants of Andrew Carnegie
That Changed America* (2012)

Edward Drummond Libbey, American Glassmaker (2011)

Henry Clay Frick: The Life of the Perfect Capitalist (2010)

H.J. Heinz: A Biography (2009)

*The Metallurgic Age: The Victorian Flowering
of Invention and Industrial Science* (2006)

Aluminum in America
A History

QUENTIN R. SKRABEC

McFarland & Company, Inc., Publishers
Jefferson, North Carolina

ISBN (print) 978-0-7864-9955-7
ISBN (ebook) 978-1-4766-2564-5

LIBRARY OF CONGRESS CATALOGUING DATA ARE AVAILABLE

BRITISH LIBRARY CATALOGUING DATA ARE AVAILABLE

© 2017 Quentin R. Skrabec. All rights reserved

No part of this book may be reproduced or transmitted in any form or by any means, electronic or mechanical, including photocopying or recording, or by any information storage and retrieval system, without permission in writing from the publisher.

Front cover images © 2017 iStock

Printed in the United States of America

*McFarland & Company, Inc., Publishers
Box 611, Jefferson, North Carolina 28640
www.mcfarlandpub.com*

To the patroness of the United States,
Our Lady of the Immaculate Conception,
that she might bring forth
an industrial Joan of Arc for the nation.

Table of Contents

Preface	1
1. Prelude	5
2. Confluence—Dynamos, Electric Furnaces and the Aluminum Process	17
3. Chemistry and Capitalism	26
4. Commercialization	36
5. Litigation and Growth	50
6. The Aluminum Industry Comes of Age	56
7. Metallurgical Wars and Monopoly	63
8. New Threats and New Markets	79
9. Birth of Alcan and the Rise of an International Industry	95
10. Paternal Capitalism and Colonialism	104
11. The 1930s Unionization and the End of Paternalism	113
12. The Great Antitrust Case	124
13. The War and the New Aluminum Industry	134
14. Oligopoly and the Dawning of the Golden Age for Aluminum	147
15. The 1950s—The Products of the 1930s Come of Age	157
16. The 1950s—The Aluminum Age	174
17. The 1960s and 1970s—The Aluminum Age	186
18. "Tin Cans" and Space Capsules	192

19.	Space Age Products	197
20.	The Best and Worst of Times, 1970–2000	203
21.	The World Aluminum Wars	210
22.	The Future, Science Fiction and New Uses	217

Chapter Notes — 229
Bibliography — 235
Index — 239

Preface

MY FIRST ENCOUNTER WITH ALUMINUM was at the Aluminum Company of America (ALCOA) Building (now the Regional Enterprise Building) in Pittsburgh. It is a 410-foot, 30-story skyscraper. When I was a young boy, my grandfather, Louis "Pap" Skrabec, took me on one of his weekly visits to downtown Pittsburgh. These trips were always historical tours of the city and its industry. Pap was a proud steelworker, but also a proud Pittsburgher and American. He was delighted as he went over the details of this aluminum landmark. The ALCOA building was the talk of the town, long before the United States Steel Building. It was designed to be the headquarters of the ALCOA, and its construction began in 1951 and was completed in 1953. The construction of the ALCOA Building was a revolutionary use of aluminum in architecture and building. The unique aluminum walls of the building are only ⅛ inch thick, giving the building a lightweight and economical design. The entire facade of the tower is sheathed in stamped aluminum panels. The windows and frames, heating and ventilating ducts, water piping and wiring system were all made of aluminum. Another innovation in skyscraper construction was the aluminum windows, which rotate to be washed from the inside. The silvery ALCOA Building was another magnificent addition to the newly emerging Pittsburgh skyline in the 1950s, and another historic industrial achievement for the city. Across the street was the rebuild of Pittsburgh's oldest church, finished in 1926 with its notable aluminum spire. Few people realize the "Steel City" might just have been called the "Aluminum City."

These tours came as my grandfather introduced me to Jules Verne and his novels. Verne had anticipated the emergence of an "aluminum city" in several of his novels. Verne would also envision the many space-related applications of the metal. It was captivating for a young boy who was bombarded by the daily news of the Space Race. As my love of science developed, I purchased my first and most treasured book, *Chemistry Creates a New World*, at Pittsburgh's large bookstore located in the basement of Kaufmann's Department Store.[1] It was in this book that I first read about the Hall Process of aluminum making, and that brought a lifelong fascination with metallurgy.

In 1964, my father packed up all the kids at the Jersey shore and took a side trip to the New York World's Fair. It was there that I saw the great aluminum geodesic dome of Buckminster Fuller. The geodesic dome was composed of 1,250 interconnected pieces of aluminum tubing, weatherproof vinyl lines on the inside, and no internal supports to obstruct the view. This dome is now used as an aviary by the Queens Zoo in Flushing Meadows Corona Park. There was also the Tower of Light made of aluminum, and endless other exhibits of the metal. The New York World's Fair would be one of the last great fairs where American companies and technology would so dominate. While at the fair, I decided to be an engineer.

The idea that aluminum represented the progress of man was demonstrated when the *Pioneer* spacecraft would carry an aluminum message from mankind to extraterrestrial life. The idea was that of the famous astronomer Carl Sagan. The first aluminum plaque was launched with *Pioneer 10* on March 2, 1972, and the second followed with *Pioneer 11* on April 5, 1973. The *Pioneer 10* and *11* spacecraft were the first human-built objects to achieve escape velocity from the solar system.

Another personal experience with aluminum came in pursuit of my metallurgy and materials engineering degree at the University of Michigan. This was in the development of the first all-aluminum engine block for the Chevrolet Vega. The Reynolds Metal Co. developed a special aluminum alloy called A-390, composed of 77 percent aluminum, 17 percent silicon, 4 percent copper, 1 percent iron, and traces of phosphorus, zinc, manganese, and titanium. The A-390 alloy was suitable for faster production die-casting, which made the Vega block less expensive to manufacture than other aluminum blocks. The university, in the earlier 1970s, was working on fine-tuning the alloy and the novel die-casting method. While my work in connection with this project was minor, I was quite proud of the result, and my first car was a Chevy Vega. It would, however, bring me many headaches and ultimately led me to be a Ford owner for the next 20 years. My projects as an engineer at National Steel, a producer of aluminum and steel, included work on studies of both metals for can making.

In writing my series of books dealing with the pantheon of great American industrialists, I maintained the perspective of an operations manager. Even more than that, I focused on the best of American industrialism. It was the combination of industrialism and patriotism that most inspired me. I purposely left out men like J.P. Morgan, who lacked that spark of patriotism and nationalism often found in America's industrialists. These early industrialists were not saints, but even with all their faults, they tended to love America. The story of aluminum holds the similar characteristics of nationalism and industrialism. Sometimes, even in highly ambitious men, you find ones of high character, such as Charles Martin Hall. More commonly, these capitalists are like each of us, some good and some bad. Andrew Mellon, with all his faults, had helped in building an improved workplace. ALCOA, like its founding fathers, had both a bright and a dark side. I hope to tell both sides.

As I was preparing this book, my daughter Diane took my wife and me to the First Center for the Visual Arts in Nashville, Tennessee, which is a beautiful example of aluminum and Art Deco. This would inspire me to consider a new angle of the use of aluminum in the promotion of culture.

As of a little more than a century ago, aluminum, while the earth's most abundant metal, remained inaccessible to man until the dawn of energy production on a massive scale. It started as the world's most precious metal in the 1850s and progressed to the world's most strategic industrial metal. While iron and steel won World War I, aluminum was the metal of the victors in World War II. Finally it would make space travel possible. The story of that advance is one of metallurgy, engineering, creativity, business, corruption, politics and the advance of civilization itself.

CHAPTER 1

Prelude

ALUMINUM HAS AMAZING PHYSICAL PROPERTIES, but its real legacy is its role in human history. It is a symbol of national power and strength even more so than gold. Aluminum stands out among all metals as a motivator. No other metal—and with the possible exception of oil—no other commodity has produced as many millionaires and billionaires. Aluminum is not found in nature; it requires man to unleash it from its ore. It is, however, a symbiotic relationship in which aluminum has allowed the advance of civilization beyond our planet. Yet aluminum is one of the world's newer commercial metals.

The name aluminum is derived from the ancient name for alum (potassium aluminum sulfate), which was alumen (Latin, meaning bitter salt). Aluminum is the most abundant metal in the Earth's crust. It is the third most abundant element after oxygen and silicon. As a metal, it is the most used non-ferrous metal in the world, and aluminum represents the second largest metals market in the world. Still, there is no naturally occurring aluminum anywhere on the planet. It is found in all clays, but its rich commercial ore is called bauxite (named for the French town of Les Baux). Large reserves of bauxite are found in Australia, Brazil, Guinea, Jamaica, Russia, and the United States. One of the largest producers of aluminum metal is the United States; states that produce the most aluminum are Montana, Oregon, Washington, Kentucky, North Carolina, South Carolina, and Tennessee. China has just recently overtaken the United States in the production and consumption of aluminum, which exemplifies China's rise in the world economy. Aluminum requires the combination of natural resources and technology. Having aluminum ore does not assure that a country will be a major aluminum producer.

The problem with aluminum ore is it occurs in one of the Earth's most stable oxides in clay deposits, but few at present are rich enough to be commercial. Even if you are blessed with rich bauxite, it requires vast amounts of energy to economically separate aluminum from its oxide ore. First, the bauxite contains only 40 percent alumina, which must be purified to 99 percent plus alumina. Then, aluminum must be separated from the refined alumina. For the production of one pound of aluminum, it takes the equivalent electricity for lighting a 40-watt bulb for 10 days. Today, a full five percent of America's electricity goes to the production of aluminum from its ore. It takes three times more electricity to make a ton of aluminum from raw materials than a ton of steel. This is why ore is not the most important natural resource, but energy is. Aluminum is often called frozen energy. Thankfully only a small amount of energy is needed to recycle aluminum because of its low melting point. Still, even today, the engineering demand for aluminum far exceeds the ability to produce it, making it the metal of dreams. Currently, aluminum represents the second largest metal market in the world after iron. A total of about 29 million tons of aluminum is needed to meet worldwide demand each year. About 22 million tons of the totals are new aluminum, and 7 million tons is aluminum scrap that is recycled for reuse.

Aluminum metal is a silver-like metal with a slightly bluish tint. It has a melting point of 1,220°F and a boiling point of 4,221–4,442°F. The density is 2.708 grams per cubic centimeter. Aluminum is both ductile and malleable. Ductile means capable of being pulled into thin wires, while malleable means capable of being hammered into thin sheets. Aluminum is an excellent conductor of electricity. Silver and copper are better conductors than aluminum but are much more expensive. Aluminum has one very useful property. When exposed to air, it combines slowly with oxygen to form aluminum oxide. The aluminum oxide forms a very thin, whitish coating on the aluminum metal. This coating prevents the metal from reacting further with oxygen and protects the metal from corrosion or rusting, which is common with other metals.

Aluminum oxide is a much different material from pure aluminum. It is an insulator, extremely hard and brittle. In nature, aluminum oxide is mainly found in rich clay ores called bauxite. Bauxite is only about half aluminum oxide. Aluminum oxide is found in some pure crystalline forms. Corundum is the most common naturally occurring crystalline form of aluminum oxide. Rubies and sapphires are gem-quality forms of corundum, which owe their characteristic colors to trace impurities. While alumina has many uses, its main use is in the production of aluminum.

The largest single use of aluminum today is in the transportation industry at 28 percent. Car and truck manufacturers prefer aluminum because of its lightweight and resultant fuel savings. Aluminum packaging, such as beverage cans and foil, make up 23 percent, and aluminum in building and construction amounts to 14 percent. Window and door frames, screens, roofing and siding, as well as the construction of mobile homes and buildings, rely on aluminum.

The remaining 35 percent of aluminum is used for a staggering range of products, including electrical wires, appliances, automobile engines, heating and cooling systems, vacuum cleaners, kitchen utensils, garden furniture, heavy machinery, and chemical equipment.

Few metals have captured the imagination of as many artists and designers as aluminum. Aluminum ranks with silver, copper, and gold as a favorite amongst jewelers and artists because it's malleable, easy to cast, and easily engraved. Aluminum can be cast into precise shapes or hammered into any form. Its earliest occurrence in workable amounts inspired jewelers to form it into pieces of art. Artists then followed with castings. It can be sand-cast or precision die-cast and can be worked on with simple tools such as files, saws, and hammers. Aluminum can take on the shiny luster of stainless steel or the dull luster of pewter. Additionally, it resists corrosion better than silver and copper. Even today, it is still an artistic metal found in giftware and art pieces.

Aluminum and its alloys offer a strange mix of properties compared to other metals and their alloys such as steel, cast iron, copper, magnesium, etc. No other metal has the range of aluminum in alloying properties. It can be a great conductor or a great insulator. It can be used to collect heat, as in cooking, or to dissipate it, as in car radiators. It can absorb or reflect light. It can be malleable like copper or hard like steel. It can take on colors. It does not corrode like steel, iron, and copper. It can be made into anything from foil to structural beams. It has been the metal of kings and the metal of the poor. It is one of the most recyclable metals, and no other metal represents the rise of mankind better than aluminum. Unlike other metals, it required a confluence of technological advances to unlock it from its ore and conquer its alloying properties.

Aluminum, as we shall see, is not without an Achilles heel, which is the potential of catastrophic structural failure called metal fatigue. Metal fatigue is the progressive and localized structural damage that occurs when a material is subjected to cyclic loadings. These cyclic loadings can be simple vibration in airplanes or ocean waves on ships. The concept of fatigue is very simple: when a motion is repeated or cyclic, the object that is doing the work, or receiving it, becomes weak. Small cracks form and eventually this leads to total failure of the part. Early and famous fatigue failures were seen in de Havilland Comet aircraft, the world's first commercial jet, which broke up mysteriously in midair in 1953 and 1954. At the time, very little was understood of metal fatigue; but it would later play a key role in the history of modern aluminum. Aluminum also requires special fabrication and welding techniques that come easily to competing metals such as steel, but require advanced techniques for aluminum.

The first industrial revolution was that of iron and eventually steel. Iron, like copper, bronze (an alloy of copper and tin), and stone, defined an age of man. Copper could be found in pure forms or in ores easily reduced by man. Iron required technology, but fairly primitive people were able to produce iron artifacts. Steel was an iron alloy which required science and technology to produce it. Still, 18th-century man was able to produce it in limited amounts, while 19th-century man conquered the technology. Aluminum had to wait for 20th-century man to conquer it because its production required science unknown to man until the late 1800s.

For many, the economic production of aluminum was a "second industrial revolution." Unlike earlier ages of copper, bronze, iron and steel, aluminum required a new source of energy to unlock it. The appearance of commercial aluminum opened up air and space travel. Aluminum clearly marks a major advancement from the beginnings of the Iron Age industrial revolution. Some believe aluminum required a new form of capitalism to make it a useful product. Famous engineer Buckminster Fuller called this "metal cartels capitalism," and it was beyond the laws and restrictions of any one country.[1] Fuller also characterized the aluminum cartel as "world masters of a line of supply."[2] In 1904, Jules Verne in his novel *Master of the World* predicted the mastery of aluminum to produce aircraft would create a master of the world. The *Apollo* moon capsule and rocket were aluminum, just as Verne predicted in his 1865 novel *From the Earth to the Moon*. The airframe of a typical modern commercial transport aircraft today is 80 percent aluminum by weight. By the 1930s, aluminum would define the real masters of the world.

Aluminum's properties make it uniquely suited for engineering applications in aircraft, spacecraft, and land transportation. It is 2.5 times lighter than steel and, with alloying, it can approach the strength of steel. Its weight advantage and strength make it popular in all modes of transportation. Today, we see it in everything from racehorse shoes to modern jet skis, and ultimately, it remains the heart of modern aviation. The energy crisis is making aluminum, once again, an alternative for car skins over steel. Aluminum is roughly ⅓ the weight of steel alloys, making it a major component in reduced fuel usage and "green" applications. It has outstanding electrical conductivity, making it extremely popular in many electrical applications. Copper, being the preferred conductor over short distances, has a disadvantage of weight in long-distance transmission, whereas aluminum is half the weight of copper. Aluminum, by the late 1890s, was making inroads into long-distance transmission lines; and today, nearly 90 percent of our high-power lines are aluminum. These long-distance electrical cables make full use of aluminum's conductivity, light weight, and strength. Similarly, its heat conductivity makes it a popular material for cooking utensils. It is exceptionally corrosion-resistant and can be used without constant painting. The dreams of its future use continue in the development of aluminum foam used in the space shuttle. Aluminum's future remains bright as new applications in sports, auto manufacturing, space, I-watches, and even transparent armor are continually coming to market.

For centuries, aluminum was the most prized of precious metals. The early appearance of this unique metal goes back to ancient Rome and China. There is a legend attributed to Roman historian Pliny the Elder that aluminum was offered to Emperor Tiberius by an alchemist. Tiberius, fearing the devaluation of his gold and silver, had the alchemist beheaded and his workshop destroyed.[3] It would seem unlikely such a legend would be true, based on the historical struggle to unlock aluminum from its ore. The required energy represents as much as 50 percent of the production cost of aluminum. In fact, the energy needed to unlock large quantities of aluminum did not become available until the 1890s, although it was possible to produce precious ounces with chemical reagents. This might explain the finding of small amounts of aluminum in the 3rd-century tomb of Chinese emperor Chou-Chu.[4] Thus, the story of aluminum is really the story of the ascent of modern man and technology. It is also a story of the confluence of evolving scientific progress, resulting in the discovery being made by two different inventors (Charles Hall and Paul Heroult) on two different continents.

The quest for aluminum was a 60-year struggle marked by progressive steps to reach the goal. The problem, as with all metals, was the separation of the metal from its natural oxide or other chemical ore. Iron requires the use of carbon and heat to wrestle it from its natural oxide ore. Aluminum oxide, however, is one of nature's strongest bonds, and carbon cannot break these bonds as it does with iron oxide. In fact, it is the thin oxide surface on aluminum that makes aluminum metal corrosion resistant. Today, we even thicken that oxide layer in a process known as anodizing to make the aluminum perform as an inert metal in the most corrosive environments. Aluminum oxide or complex oxides are the only ores of aluminum. Many metals such as copper even exist in their pure state in nature or in ores of soluble salts. Metallic salts, such as copper salts, could be created to produce pure copper using an electrical current and passing it through a water solution of copper sulfate. The problem with aluminum oxide ore is that it is not dissoluble in water. Thus, the problem became how to prepare a low-temperature solution of aluminum to allow a current to pass through. That quest would start with the famous chemists at the beginning of the 19th century.

It was not until mid–19th century that elemental metallic aluminum was produced in even small quantities. While an aluminum-iron alloy had previously been developed in 1820 by British scientist and inventor Humphry Davy, it lacked the purity to be called metallic aluminum. Humphry Davy also started experiments in electrolysis in 1800, but he struggled to isolate metals by putting a current through solutions of their alkali salts, which did nothing more than free hydrogen. However, he had much better results when he started to electrolyze heated molten compounds, first isolating potassium from potash and sodium from table salt in 1807. The following year, Davy used electrolysis to produce elemental calcium, strontium, barium, and magnesium, which led to the identification and naming of aluminum. Aluminum (al-oó-min-um) was the name given by Davy. Sir Humphry made a mess of naming and spelling this new element, first spelling it alumium in 1807, then changing it to aluminum, and finally settling on aluminium in 1812. The name "aluminium" (al-yoo-miń -iuhm) was used in England and France by popular writers, such as Charles Dickens and Jules Verne, which made "aluminium" the acceptable name in Europe.

In order for the reader to better understand country names for it, a brief discussion of the spelling is needed. The debate over the name persisted for some time in the United States. *Noah Webster's Dictionary* of 1828 includes only aluminum, though the standard spelling among U.S. chemists throughout most of the 19th century was aluminium. During the 1890s, the preferred spelling in the United States became aluminum. The origin of the American spelling was seen as a misspelling in one patent of American inventor Charles Hall. The issue may have been settled in 1907 when ALCOA, the world's largest aluminum company, used the aluminum spelling. However, in Hall's and Heroult's original patents, the word was spelled "aluminium"; but the American Chemical Society debated the name in 1925 and decided to stick with aluminum.[5]

Regardless of spelling, the quest for elemental aluminum began in Europe in the early 1800s. The first breakthrough to the unlocking of aluminum from clay came in 1825. That same year, Danish chemist Hans Christian Oersted heated aluminum chloride with a potassium amalgam metal to produce impure bits of aluminum metal at an estimated cost of $60 per ounce. Gold sold at $25 an ounce in 1825, and the average worker during that time made $600 in 3 to 4 years of work. Oersted's method allegedly produced aluminum with too many impurities for him to be able to take full credit in history for isolating elemental aluminum.

German chemist Fredrick Wöhler improved the process to the point that he was able to produce pinhead quantities of pure aluminum by the 1840s, thus becoming the first researcher to claim to have created elemental aluminum. Using electrolysis, the German Robert Wilhelm von Bunsen, who coincidentally had taken Wöhler's place as a chemistry teacher at the Higher Polytechnic School at Kassel in 1836, further improved the purity. However, typical of so much of the confusion surrounding the story of aluminum, Oersted was proven to have been the first after all. The proof would come a hundred years later in 1936, when scientists at the Aluminum Company of America followed a lost journal paper of Oersted using his directions. The result was aluminum of elemental purity, and the experimenters called for "the world to atone for injustice by giving the Dane [Oersted] credit for the discovery of aluminum."[6] Historical credit is hard to change, but many books now say Oersted was "probably" the first. Danish King Christian X used an aluminum crown to make the point in the 1930s.

While the Wöhler and Bunsen process greatly improved the quantities and purity, it did little to bring down the cost initially. It was 1850 when they were able to achieve a cost of about $40 an ounce, which was still very pricey. The real cost breakthrough was achieved by French chemist Henri Sainte-Claire Deville (1818–1881) as he used a new process in the 1850s. Sainte-Claire Deville used a special type of clay found in France, near the village of Les Baux. The clay was high in aluminum content and became known as bauxite, named after the village. While any clay can be considered aluminum, bauxite offers major economic advantages for the commercial production of aluminum. Bauxite is a weathered form of clay having aluminum hydrates, and it can be readily converted to pure aluminum oxide (alumina).

Sainte-Claire Deville used sodium instead of potassium to separate aluminum from the oxide. Because sodium was less expensive and easier to obtain than potassium, Deville was able to produce more aluminum—enough to make an ingot. Deville drove the price down in 1854 to $16 a pound by the patronage of Napoleon III, who financed Deville's research. While this lower cost allowed new possibilities for aluminum, it was still too expensive for common use, but the combination of light weight and strength offered a vision of many potential uses to inspire others.

One of those technical visionaries was Louis-Napoléon Bonaparte (1808–1873), the first president of the French Second Republic. He was the nephew and heir of Napoleon I and would be called Napoleon III, the Emperor of the Second French Empire. While he was the first president of France to be elected by a direct popular vote, he was blocked by the Constitution and Parliament from running for a second term. He organized a coup d'état in 1851, and then took the throne as Napoleon III on December 2, 1852, the 48th anniversary of Napoleon I's coronation. His political irregularities aside, Napoleon III was a true visionary who played a key role in the development of aluminum.

Napoleon III was a prince of technology, rivaling Queen Victoria's husband, Prince Albert, in the Victorian era. Napoleon III promoted public works, the construction of railroads, and other means of furthering industry and agriculture. He also took a personal interest in the rebuilding of modern Paris and was an ardent supporter of French inventors. He believed it was the role of government to promote economic growth and technology. Napoleon III offered many prizes to inventors for the development of steel and aluminum. He had offered various awards in the late 1840s that attracted chemists, like Sainte-Claire Deville, from all over the world. Napoleon III financed Deville to continue his aluminum experiments on a larger scale. The work was performed on at the Javel Chemical Works, and while the process appeared practical, it was expensive. Despite the cost issue, Deville was soon manufacturing two tons of aluminum each year. Over the next 20 years, technical improvements increased production to 10 tons per year, with the cost remaining around $16 a pound.

Aluminum proved to be a special fascination for Napoleon III. For years, Napoleon III had been a promoter of high technology, and even prior to Deville's work, he saw a bright future for this metal and did more than anyone to promote it. In the early 1850s, he had a set of aluminum tableware produced for his most important guests, while lesser guests ate on gold. He also had a suit made with aluminum buttons, had a baby rattle made of gold and aluminum for his son, and had the brass eagles on the flagstaffs of the imperial guard cast in aluminum. As the price came down for aluminum, with his support of Deville's research, he proposed suits of armor for his military guard and had a prototype of the helmets produced. Napoleon III made aluminum part of French culture, and soon the wealthy French replaced gold foil with aluminum foil for wall decorations. Additionally, fashionable and wealthy French women began wearing jewelry crafted of aluminum made by the famous goldsmith Christofle. Christofle, at the request of Napoleon III, crafted an aluminum watch chain for the King of Siam. Other European royalty followed his example, using aluminum crafted by their own jewelers. Napoleon foresaw the use of aluminum horseshoes to improve the speed of his cavalry. At the Paris Exposition of 1855, Napoleon III exposed the world to the future of this metal.

The story of aluminum is one of visionaries, like Napoleon III, who foresaw the potential of this metal before the technology existed to produce it economically. It was these visionaries who imagined aluminum's uses in transportation and building. Napoleon III had supported a major exhibit at the 1855 Paris World's Fair, highlighting Deville's "aluminum from clay" process. While this exhibit only had a small amount of aluminum, it offered a giant vision for the metal. It would inspire the scientific research needed in order to make aluminum a commonplace metal. The exhibit caught the attention of three great writers—Charles Dickens of England, Nikolai Chernyshevsky of Russia, and Jules Verne of France, who ignited the imagination of the world. Ultimately, the exhibit was the root of commercialization of aluminum.

The general public was also intrigued. Charles Dickens, commenting on

Deville's initial success in 1855, wrote: "Aluminum may probably send tin to the right about face, drive copper saucepans into penal servitude, and blow up German-silver sky-high into nothing."[7] Russia's Chernyshevsky saw aluminum as the "metal of socialism" after seeing the "aluminum from clay" exhibit. In his 1863 novel, *What is to be Done?*, he predicted: "How light is the architecture of this inner house, how small the piers between windows ... but what are these floors made of? And the frames of these doors and windows? Silver? Platinum?... sooner or later, aluminum will replace wood and even stone. How rich it looks."[8] It would appropriately be a French author who showed the potential of aluminum to the world—a young Jules Verne, present at Napoleon III's World Fair exhibit, foresaw its great future.

Production worldwide in 1869 averaged about 2 tons. Jules Verne would bring demand of aluminum to the world in his many scientific tales. In Verne's *From the Earth to the Moon* (1867), he predicted the first commercial use of aluminum. Verne's novel required the manufacturing of a 20,000-pound space capsule to take men to the moon. Amazingly, it would be the same weight of the *Apollo 8* capsule that orbited the moon more than 100 years later. At the time of Verne's writing in 1867, aluminum cost $15 a pound in Europe, which had just been reduced from $16 a pound with France's first aluminum production plant. Still, $15 a pound was an extremely high price, and Verne's fictional order of 20,000 pounds to build a space capsule was far more than the world's total inventory at the time. It would be another 20 years before such an order could possibly be filled. Yet, Verne realized aluminum would be necessary to put the moon trip into the realm of reality. In the next few years, Verne's novels would feature aluminum airplanes, skyscrapers, ships, helicopters, and submarines. His novels also included accurate descriptions of aluminum metallurgy and science, which inspired generations of engineers to advance the application of aluminum. A few years later, British science fiction writer H.G. Wells used it for his fictional Martian spacecraft in *The War of the Worlds*. Both Wells and Verne were extremely popular writers in America.

In 1867, in his novel titled *From the Earth to the Moon*, Verne characterized aluminum: "Unquestionably, my friends. This valuable metal possesses the whiteness of silver, the indestructibility of gold, the tenacity of iron, the fusibility of copper, the lightness of glass. It is easily wrought, is very widely distributed, forming the base of most of the rocks, is three times lighter than iron, and seems to have been created for the express purpose of furnishing us with the material for our projectile."[9] Just over 100 years later, there would be an aluminum capsule on the moon.

Aluminum would first make headlines in America in 1884 with a 100-ounce aluminum cap on the completed Washington Monument. By 1884, world aluminum production had only increased to 4 metric tons, a small amount compared with the 3,000 metric tons of silver that were produced that year. Only 112 pounds of aluminum were produced in the United States in 1885, and virtually all of it was produced by a Philadelphia immigrant named William Frishmuth (1830–1893). The bulk of the remaining amount came from France, Germany, and England. The American process of Frishmuth's method was a two-stage chemical process lacking the efficiency of the electrolysis/chemical methods being used in France. Frishmuth's aluminum cost between $16 and $20 a pound, while the French price was $12 a pound.

In 1876, Frishmuth produced the first authenticated aluminum castings made in America at his Philadelphia foundry. Frishmuth would be commissioned in 1884 to produce the aluminum cap for the Washington Monument. His annual capacity for aluminum production was less than 200 pounds per year by 1885. Frishmuth studied with Wohler in Germany before coming to America, and he used a solely chemical process, unlike the electrolytic processes used today. The foundry was declared an ASM (American Society for Materials) Historical Landmark in 1985. While Frishmuth's process lacked the productivity of his French counterparts, he pioneered the first methods of casting the metal. Frishmuth had exhibited his first castings at the Philadelphia World's Fair in 1876. Frishmuth's aluminum cap of the Washington Monument made world headlines in 1884 and bring the wonder of this metal to new heights.

The Washington Monument had actually been started in the 1850s, but was put on hold during the Civil War. Because Frishmuth had previously done plating work for the Washington Monument and the Army Corps of Engineers, and also had been a friend of President Lincoln,

Cast Aluminum cap on Washington Monument with lighting rods (Library of Congress).

the government asked him to construct a small metal form for the top of the monument. The small pyramid on top of the monument was to be artistic and function as a lightning rod. Frishmuth suggested aluminum; and in 1884, he cast the cap, which was the first architectural use of aluminum in the world. At that time, the cost of aluminum was $16 a pound. For perspective, the cost of silver was $19 a pound. Before the 6-pound aluminum pyramid was installed on top of the monument on December 6, 1884, it was on display at Tiffany's jewelry store in New York, and people lined up for blocks just to get a look at the unique casting. At the very top, the words "Laus Deo" or "praise be to God" are engraved. Frishmuth's casting techniques would change the typical metallurgist's view of aluminum as a difficult metal to cast. Afterwards, even cast aluminum dentures were made in 1886. However, the problem of cost inhibited the use of aluminum. Kaiser Aluminum commissioned the replica of the aluminum pyramid—cast 100 years later in the same Philadelphia foundry—and gave it to Oberlin College in 1984.

Chemists and metallurgists around the world continued to look for cheaper alternatives for aluminum production. In 1886, an American, Alexander Castner, discovered a more economical method of producing sodium, and this reduced the cost to make aluminum to about $8 per pound; but this cost was still too high for mass consumption. Moreover, Castner lacked capital to build a large processing plant like those in France. The Deville process would remain the dominant method until the later 1890s, while Deville continued to improve on his casting techniques. The cast statue of Eros in Piccadilly Circus was a Deville casting made in 1892.

The Deville process used carbon-based heat to drive chemical reactions, and Deville also used carbon to power his electrical dynamos as an electrical source. Because this method of carbon heating and separate electrolysis was expensive, even with economy of scale, Deville was hard pressed to break the $15-a-pound mark. The breakthrough came with the dual use of electricity both to heat and provide electrolysis of molten salts. This revolutionary process would come in 1886, but it would take several years for complete commercialization. Part of the issue would be the development of cheap electricity that would come with the power station at Niagara Falls and the availability of alternating current. In 1897, all the engineering and science would fall into place for the production of lower-priced aluminum.

Inexpensive aluminum would really define the start of a second industrial revolution. By 1897, the price would fall to 54 cents a pound, resulting in the commencement of the aluminum age. This would inspire even more unusual applications, like an aluminum dress that was made in 1898 for a woman in Queen Victoria's court. Also in 1898, Pope Leo XIII had the dome of the Church of San Gioacchino coated with aluminum; the aluminum dome remains in excellent condition today. Aluminum bicycles and auto parts appeared in the 1890s. The Navy began experimenting with ship applications in 1898 while the Army started using aluminum canteens. The first sports car featuring a body made of aluminum was presented to the general public at the Berlin International Motor Show in 1899. Soon after, aluminum would even play a role in early flight applications. And thus, the story of how we got to low-cost practical applications of aluminum is really the story of modern man and the advance of technology in general.

CHAPTER 2

Confluence—Dynamos, Electric Furnaces and the Aluminum Process

IN 1886, AMERICAN CHARLES MARTIN HALL (1863–1914) and Frenchman Paul Heroult (1863–1914) would develop almost the identical revolutionary process at the same time. This is not surprising in the history of invention, which has seen many simultaneous discoveries. Science and engineering principles evolve through the work of many over time. Still, history, or at least general history, has always preferred to give credit to an individual. For many, it was a matter of who filed for the patent first, as it was with Alexander Graham Bell, who beat Elisha Gray by an hour to patent the telephone (an event that would be part of aluminum history). Another example was when Thomas Kelly filed for a patent around the same time as Henry Bessemer for a commercial steelmaking process. Bessemer won the patent and attracted publicity by his aggressive commercialization of the steel process. Then, 30 years later, the Supreme Court correctly gave the patent and credit to Thomas Kelly as he lay on his deathbed. However, it was too late for the history books because the Bessemer name was everywhere, including the naming of towns and cities, and in the end, Kelly's name was lost to mainstream history. Nikola Tesla filed first to patent radio, but was turned down in preference to Guglielmo Marconi. Marconi received the first patent for radio in 1896 and, through aggressive commercialization, won the historical credit. In 1911, Marconi won the Nobel Prize, and Tesla was furious. Tesla sued the Marconi Company for infringement in 1915, but was in no financial condition to litigate a case against a major corporation. It wasn't until 1943—a few months after Tesla's death—that the U.S. Supreme Court upheld Tesla's radio patent. For Tesla, however, it was too late for the historical credit he deserved since generations of schoolchildren had been taught that Marconi was the inventor of radio.

Invention, however, is more often a confluence of knowledge and science with many steps and preceding advances than that of the lone scientist. Invention might be better called evolutionary versus revolutionary, and aluminum is no different. Commercial aluminum would require the evolution and development of electricity from dynamos (versus batteries), and ultimately, the development of the electric furnace. The science and chemistry of aluminum had advanced rapidly in the early 1880s, but the needed engineering was lagging. The scientific community was rapidly coming to the idea that both electrically generated heat and electrolysis could be used to commercialize aluminum production. Battery-supplied electricity of the period lacked the power for commercial aluminum production. The first engineering advance to make the Hall-Heroult process commercial was the electric dynamo, which eliminated the need for endless rows of lead-acid batteries to supply electricity. The dynamo used mechanical power from water or steam to produce electricity. The modern dynamo, fit for use in industrial applications, was invented independently by Sir Charles Wheatstone, Werner von Siemens, and Samuel Alfred Varley. It was not until 1876 that commercial dynamos would be available, and it would be Siemens's dynamo that powered the first electrical arc lights. The next evolutionary step was made by a University of Michigan engineer, Charles Brush, who founded Brush Electric in Cleveland, Ohio. Brush would build the biggest dynamos to power huge demands of commercial electric furnaces and arc lighting. Brush dynamos and arc lights were used on the streets of Cleveland before Edison's incandescent system. On April 29, 1879, they were used to light a public park in Cleveland.

The final step in the Hall-Heroult process was made by the forgotten Cowles Brothers. The Cowles Brothers used electric furnaces to drive their thermal chemical processes to produce alloys of copper and aluminum. It was known as a carbothermic process.[1] These new electric furnaces were built in 1884 using Brush dynamos, which at the time were the largest in the world. The Cowles process would lead to the future Hall-Heroult process that used electricity to heat molten salts and dissolved alumina, followed by electrolysis to obtain pure aluminum. But the first method of the Cowles Brothers was to use arc electric melting to drive chemical reactions to make intermediate aluminum alloys.

Eugene H. Cowles (1855–1892) and Alfred H. Cowles (1858–1929), sons of newspaper publisher Edwin Cowles of Cleveland, Ohio, built high-temperature electric furnaces during the late 1880s in Lockport, New York. The Cowles Electric Smelting and Aluminum Company was headquartered in Cleveland and was a pioneer in electrochemical processes, as well as the use of electricity in industrial and chemical processes. In 1885, the Cowles Brothers started to make a copper alloy known as aluminum bronze which had 10 to 20 percent aluminum and was produced by an electrothermic process. This was the forerunner of the electrolytic process of Hall and Heroult, used to produce pure aluminum. The Cowles Brothers generated electricity by using Erie Canal turbine power to produce current for their dynamos. Early experiments of Charles Hall had used direct current from acid-lead batteries. Battery current was expensive and less powerful, but mechanical dynamo current would not be commercially available until the 1890s. Hall's experimental batteries used a pound of lead to produce an ounce of aluminum. It was also cheaper to make aluminum bronze (45 cents a pound) compared to pure aluminum ($8 a pound).

The Cowles method of making aluminum bronze consisted of a firebrick wall furnace with two carbon electrodes. The furnace was charged with pure copper, alumina and charcoal, and the electrodes were then lowered into the furnace charge. An electric current was supplied by a Brush dynamo. The charge of copper and carbon electrodes on each end of the furnace created a circuit that melted the copper and alumina mix. Copper was the key because it set up the electric circuit and allowed the flow of electricity, whereas a charge of pure alumina would have prevented the electric furnace from operating. Alumina (aluminum oxide), by itself, is an insulator, unlike elemental aluminum, which is an excellent conductor. The melting was done by the electrical arc. The charcoal in the charge was a reducing agent to free oxygen from the alumina. After an hour, the melted (molten) product (known as a "heat") was poured into a 60-pound ingot of about 20 to 40 percent aluminum. Once the chemical laboratory tested the percent analysis, the ingot was re-melted in the electric furnace, and copper was added to bring the aluminum content down to the standard of 10 percent for aluminum bronze (90 percent copper). The furnaces could produce about 300 pounds of aluminum bronze every 24 hours. The furnaces could achieve temperatures in excess of 3,000 degrees F.

In 1884, industrialists saw the future of this amazing aluminum bronze as far greater than that of pure aluminum. Aluminum bronze castings proved to be superior to regular bronze (copper and tin) in naval applications, and aluminum bronze was rapidly finding success in the market of ship propellers as well as other product markets. George Westinghouse was experimenting with adding aluminum bronze to Babbitt bearing metal at Westinghouse Machine Company. Westinghouse engineers were also using aluminum bronze in corrosion applications, such as valves, for railroad air brakes. Aluminum bronze was being used in Europe in forging and industrial hammers, too. In addition, the government and *Scientific American* were predicting a huge market in cannons, bridge construction, and ship armor made out of aluminum bronze.[2]

Both Germany and America were testing cannon and ordnance made out of aluminum bronze.

The aluminum bronze cannon would, according to engineers of the time, exceed the strength of iron-wrought cannons. Eugene Cowles predicted in an 1886 paper before the Franklin Institute that aluminum bronze would revolutionize warfare.[3] At the time, steel ordnance cost about $1 a pound, while Cowles predicted he could supply aluminum bronze ordnance at 45 cents a pound.

Interestingly, in France, on an experimental level, Paul Louis Toussaint Heroult (1863–1914) was working on his own electric furnace in the pursuit of aluminum and its alloys. Heroult had been born and nurtured in aluminum. His father had worked for an aluminum company in Salindres, France, that used the Deville process. At 15, Heroult dedicated himself to aluminum development after reading Deville's book on aluminum in 1878.[4] He inherited a small tannery from his father in 1885 which had a powerful steam engine, and he used family savings to purchase a dynamo and attached it to the steam engine. This would allow him to build his own electric furnace as well as experiment with the production of aluminum and aluminum bronze. While his electric furnace followed the development of the Cowles Brothers, it was 19 months ahead of the young American, Charles Martin Hall, in using a furnace with internal electric heat.[5] At the time, young Charles Hall was using a coal-fired furnace to melt the charge, with smaller battery-powered electrodes to perform the electrolysis of the molten charge. Ultimately, Heroult would be the inventor of electric furnace steelmaking as well as co-inventor of the aluminum process. Heroult would use the electric furnace for melting and electrolysis, and was taken off track, like the Cowles Brothers, in his quest for the direct production of aluminum bronze.

Merle Chemical in France was also making aluminum bronze. Initially, Cowles Electric had little interest in the processing of pure aluminum, since the aluminum alloys made indirectly were more profitable. The discussion of the Hall experiments in early 1886 not only offered a potentially cheaper method to make aluminum, but also aluminum bronze. Cowles Electric did start experiments on the use directly, but never fully, of high carbon with alumina to make pure aluminum. The company was able to produce some aluminum and developed its commercialization. Still, the Cowles electric furnace was a necessary step in the commercialization of aluminum. To be clear, regardless of the chemistry, the Cowles's process used electricity to generate heat to melt the furnace charge of alumina and copper. Copper provided the electrical resistance to heat the charge, and the carbon in the melted charge then reduced the aluminum oxide (alumina). Thus, the Cowles process was a thermic process chemically.

The future breakthrough of the Hall-Heroult process would use electricity for electrolysis to produce aluminum.

What was not in dispute was the credit for the invention of a commercial electric furnace, which was attributed to the Cowles Brothers by major scientific and engineering groups such as *Scientific American*, the Franklin Institute, and the *Engineering and Mining Journal*.[6] Donald Wallace, an early aluminum historian of Harvard, noted: "The success of the Cowles furnaces was followed by a broadcast of their results in the leading scientific journals of Europe and America. Literature concerning aluminum was widely disseminated by Cowles Company. In attracting world-wide attention to the potentialities of electric smelting by furnishing other workers with both knowledge and stimulus, the Cowles brothers played a leading part in the industrial development of electrochemistry."[7] In addition, the development of the Cowles electric furnace would lead to the commercialization of many metals and materials, including alloy steel, silicon, and silicon carbide. In fact, for a brief time, the Cowles Brothers worked with Charles Hall to make Hall's experiments commercial.

Interestingly, another set of brothers in England were also building electric furnaces—the Siemens brothers. There were eight of the Siemens brothers, who brought a long list of inventions to their name. Charles (1823–1883) had taken the lead in the 1850s with the invention of his open-hearth furnace along with his younger brothers, Carl (1829–1906) and August Friedrich (1826–1904). Charles William Siemens (originally Carl Wilhelm Siemens and later Sir William Siemens) would become famous for his furnaces. The open-hearth furnace that was heated by gas produced low-grade coal outside the furnace. In the 1900s, this invention that was first used in glassmaking was soon widely applied in steelmaking and eventually supplanted the earlier Bessemer process of 1856. In 1878–79, Sir William Siemens took out patents for electric furnaces of the arc type. Siemens first demonstrated the arc furnace in 1879 at the Paris Exposition by melting iron in crucibles. In this type of furnace, horizontally placed carbon electrodes produced an electric arc above the container of metal. Siemens's focus was on steelmaking and overlooked its full potential until after the commercialization of the Hall-Heroult process. The electric furnace thus offered a source of heat to melt aluminum ores and the electric current to perform electrolysis.

The development of the electric furnace would set the stage for two young men in different countries to make aluminum production a reality. The discovery of the process, however, would only be a small step towards the commercialization of aluminum; and eventually, it would lead Charles Hall to work briefly with the Cowles Brothers. Ultimately, the commercialization on a massive scale was the most efficient merging of furnace and technology, and that confluence would lead to the monopolistic giant, the Aluminum Company of America (ALCOA).

Hall and Heroult used electrolysis in their revolutionary process. Electrolysis is the passage of a direct electric current through an ion-containing solution made of water, acid, or molten salts. The liquid solution contains metal ions, which can be then turned into pure metal. Electrolysis produces chemical changes at the electrodes, which allow a pure metal to be deposited. For example, some salt ores of metal can be dissolved directly in water, such as in silver and copper. Electrolysis is the chemical process used in many common electroplating procedures. Aluminum ores cannot be dissolved in water or acid, thus requiring the use of molten salts to dissolve them for electrolysis. The story of Hall and Heroult started with their first failures in trying to dissolve aluminum ore in water, then acid, before finally, successfully arriving at using melted (molten) salts to dissolve alumina (aluminum oxide) into aluminum ion solution.

The real breakthrough that Hall and Heroult succeeded in achieving was to find an effective solvent to dissolve alumina for the electrolysis of the molten bath. While alumina was readily available in high-purity form, its high melting point had deterred its use in electrolysis. Alumina cannot be dissolved in water or melted directly into a molten liquid to perform electrolysis on, but Hall and Heroult found a way to form a molten bath of alumina. Substantial energy was still needed for heating the materials and for the electrolysis, but it is much more energy-efficient than melting the oxides themselves. Hall and Heroult found a way to produce a molten bath of alumina and aluminum ions. It was determined that cryolite, with its reasonably low melting point, allowed the reduction process to occur with the metallic aluminum sinking to the bottom of the cell. Cryolite, meaning "ice rock," was an uncommon milky white and glassy mineral found in Greenland. Originally, some chemists believed it to be a commercial ore of aluminum like alumina. In 1883, Russian chemist V.A. Tyurin found a less expensive way to produce pure aluminum by passing an electric current through a molten (melted) mixture of cryolite and sodium chloride (ordinary table salt). The aluminum content of cryolite is only 13 per cent, compared to approximately 50 per cent in bauxite. Bauxite is the major source of aluminum in the Hall-Heroult process and requires a high aluminum content to be economical. Besides the low aluminum content, cryolite is extremely rare, possibly the only mineral on earth ever to be mined to total depletion. It is presently synthesized in large quantities in order to meet today's needs. Cryolite would offer the key, but not as an ore.

The role of cryolite in aluminum production goes back to some failed experiments of Swedish chemist Jöns Jacob Berzelius (1779–1848). In 1930,

2. Confluence—Dynamos, Electric Furnaces and the Aluminum Process 23

The aluminum pots at Pittsburgh Reduction Company, 1897 (Library of Congress).

Professor Holmes, Chair of Oberlin's Chemistry Department, reported that Berzelius "heated cryolite, the double fluoride of aluminum and sodium, with potassium. Unfortunately, he used an excess of potassium and got an alloy of aluminum with potassium. Had he used an excess of cryolite, Berzelius would now have credit for presenting aluminum to science."[8] Years later, Berzelius's failures would be noted in court over patent ligations. French chemist Deville had actually used molten cryolite alone to extract a small amount of impure aluminum.[9] The famous French chemist and co-worker of Deville, Louis Le Chatelier, had filed a patent in 1861 suggesting the use of cryolite.[10] The problem was that Le Chatelier had not solved the electrical furnace arrangement to make his process yield aluminum.

The real accomplishment of the Hall and Heroult process was in the application of cryolite (sodium aluminum fluoride) as a flux and molten bath, which, used with the proper electrodes, would yield aluminum through electrolysis. Even here it may seem surprising that two independent researchers would fall on the identical solution; but the story is the same, that of converging streams of knowledge. Fluorite (calcium fluoride), the primary mineral source of fluorine, was first described in 1529 by Georgius Agricola. Agricola described fluorite minerals as additives used to lower the melting point of metals. Agricola used the Latin verb fluo meaning "flow," which became associated with it, and the name later evolved into fluorspar (still commonly used to lower melting points). Charles Hall had initially experimented with fluorspar, but found its melting point too high. Cryolite offered a much lower melting point, thus forming a liquid that could then dissolve alumina. Greenlander Dr. Edward Kleiner-Fiertz and Frenchman Adolphe Minet had tried with limited efforts, in their earlier work, to release aluminum from cryolite directly with fused cryolite baths. By the 1880s, the Russians had dissolved cryolite in molten sodium chloride, then used electrolysis to release small amounts of aluminum. Considering the small amount of aluminum in cryolite, as well as the rarity and cost of cryolite, the process was impractical and not economical. Still, science was getting close to unlocking aluminum from nature. The interest in cryolite turned from its possible usefulness as an ore to a possible fluoride flux.

In the 1880s, the study of fluorides was popular in the universities of Europe. In fact, it would be in 1886 that French chemist Henri Moissan used electrolysis of hydrofluoric acid to isolate elemental fluorine gas, and later he would win the 1906 Nobel Prize in Chemistry for this work. Both Hall and Heroult had mentors with ties to this research that would lead them to the use of cryolite. They had tried to use readily available fluorides such as fluorspar and magnesium fluoride, but failed to melt alumina because of their high pointing points. They found the perfect fluoride to dissolve alumina in cryolite. Cryolite acted as a "flux" for the Hall-Heroult process as alumina is dissolved in a bath of molten cryolite (a sodium aluminum fluoride mineral) at a temperature of about 1800 degrees F. This molten bath, or electrolyte, is held in a cell consisting of a cast iron shell lined with carbon, which serves as a cathode, and has carbon anodes suspended within it. The electrical current passing through the electrolyte separates the dissolved aluminum oxide into aluminum and oxygen. The metallic aluminum is deposited at the bottom of the cell, and the oxygen is deposited on the carbon anodes. In a 1986 centennial study of the Hall-Heroult process it was noted: "The genius of Charles Hall was his discovery that fluorides and particularly cryolite could be used as a solvent for forming a stable electrolyte by *reaction* with what he perceived to be the cheapest pure compound of aluminum, alumina."[11]

While the legal battles would combine the two processes for financial reasons, the merged processes also proved more efficient for mass production. There was an important difference in the Hall and Heroult processes to make the molten bath for electrolysis. Initially, Heroult used aluminum bronze left in the furnace to act as a conductor for his charge of alumina, using electrodes for melting and electrolysis. Heroult soon found that he could use the carbon electrode to produce the heat for melting, much like electric arc welding today or electric arc furnaces in steelmaking. Hall used external gas and coal heating and a flux (sodium fluoride) to make a molten bath, which Heroult also used as another alternative. Carbon electrodes were then used for the electrolysis, which would become common to both processes. Since sodium fluoride lowered the melting temperature of the charge, Heroult arrived later at the same point in order to reduce the energy demands.[12] Eventually, all external heating was removed from the continuous Hall-Heroult process.

The Hall-Heroult process can be summarized in the following steps:

1. The bauxite (red-brown solid), which is alumina (aluminum oxide) mixed with impurities, is extracted from the Earth.
2. The extracted aluminum oxide is then treated with alkali to remove the impurities. This results in a white solid in a pure aluminum oxide, or alumina. It takes four tons of bauxite to make two tons of alumina to then make one ton of aluminum.
3. Next, the aluminum oxide is transported to huge furnace tanks. The tanks are lined with graphite, this acts as the cathode. Blocks (electrodes) of graphite hang in the middle of the tank and act as anodes.
4. The aluminum oxide is then dissolved in molten cryolite. This lowers the melting point, which reduces the total costs of the process. The heat required to keep the mixture molten is provided from resistive heating of the molten bath (electrolyte) by the current passing through the cell. The furnace and/or cell is called a pot, thus the use of the term "pot room."
5. Electricity is passed through and electrolysis begins. Electrolysis, in this instance, is the decomposition of alumina into pure aluminum and oxygen using electricity.
6. Aluminum is deposited on the bottom of the furnace or pot, where it can be cast into ingots. To maintain full efficiency, the process operates 24 hours a day, 7 days a week.

Chapter 3

Chemistry and Capitalism

THE DISCOVERY OF A PROCESS FOR commercial aluminum would ultimately come to two different men in two different countries—Charles Martin Hall in a middle American small town, and Paul Heroult in a village in northwestern France. Both men were born in 1863 and had a passionate love for science. Both were mentored by accomplished chemists who were well versed in current literature in the field. They were, however, far more different than similar as young men. Hall was a bookish introvert, and he would essentially be considered a nerd in today's society. On the other hand, Heroult was a man of the culture, who enjoyed dating, partying, and playing billiards and other games. What brought them to the same point in history was their drive for fame and fortune. They represented the extremes of Victorian scientist personalities, but each individual was a pure capitalist. Both saw the aluminum process as a road to success.

Charles Martin Hall was born to Herman Bassett Hall and Sophronia H. Brooks on December 6, 1863, in Thompson, Ohio. His father graduated from Oberlin College in 1847 and became a Presbyterian minister. He studied for three years at the Oberlin Theological Seminary, where he met his future wife. They married in 1849, and the next ten years were spent performing missionary work in Jamaica, where their first child was born. They returned to Ohio in 1860 when the outbreak of the Civil War forced the closing of foreign missions. Herman had one brother and three sisters, one of whom died in infancy.

The Oberlin College of Hall's father's day was a socially progressive college, having admitted both women and blacks as students. It was a hotbed of abolitionism, an important part of the Underground Railroad, and also prominent in sending Christian missionaries abroad. By the 1860s, Oberlin had produced not only missionaries, but a number of foreign diplomats. Herman had a liberal arts education, including a course in chemistry. When Charles was born in 1863, the country was in the midst of the Civil War. Oberlin College would make history that year by graduating the first black women.

Charles's earliest education was at home with his mother teaching him. He became an avid reader, and his earliest heroes were Thomas Edison and George Westinghouse. He would later say that his interest in chemistry came from reading his father's college chemistry text. Much like the case of Thomas Edison, his love of chemistry led to early experimentation. Most of these were simple experiments using household chemicals, and were performed in his first laboratory—the kitchen.

Herman Hall was committed to the idea that his children would be educated at Oberlin and moved the family there in 1873. The Hall family purchased a house on 64 East College Street. Charles Martin Hall took his preparatory work in Oberlin High School, and he would often go to the college to purchase test tubes and other equipment for his attic laboratory.

Charles Hall, 1887 (courtesy of Carnegie Library of Pittsburgh).

His high school education was supplemented by one year in the Oberlin Academy, including lessons in the Conservatory of Music. He enrolled in Oberlin College in 1880, when the college had 1,057 students. The Science and Natural Philosophy Department had 320 students in majors such as chemistry, geology, mineralogy, botany, and zoology. The department was based in a three-story building known as Cabinet Hall, after the natural history cabinet on the third floor. Built in 1838, the college was lacking physically. The only available water was from the roof gathering system, coal stoves struggled to keep the classrooms warm, and pipes usually froze in the winter and required thawing by Bunsen gas burners. Nonetheless, Oberlin College was attracting some of the nation's best science teachers and students.

Oberlin had the famous inventor, contestant to the invention of the telephone, and founder of Western Electric Manufacturing Company, parent firm of Western Electric Company, Elisha Gray (1835–1901), teaching at the college in the newly founded physics department. Even before entering college, Hall had attended public talks by Gray on the invention of the telephone. (You may remember that Gray had gotten to the patent office an hour later than Bell did, as referenced in the previous chapter.)

Charles dreamed of being an inventor, as did countless other young scientists at the time. Like many ambitious boys of the period, he dreamed more of fame than money. Hall's brother noted the following: "Aside from his studies in school and his self-imposed tasks, every moment was devoted to reading or studying along scientific lines, with occasional relaxation in music. His mind turned already to invention, and his college days were filled with dreams of discovery which should bless the world."[1] Hall was a conservative invert who opposed smoking, drinking, and modern dancing. Later in life, he actually considered withholding his bequest to Oberlin, when they thought about lifting a ban on smoking.[2] He would have a college sweetheart, but he never married. His true love was chemicals and laboratory work.

Hall spent his high school days and first year of college in search of a project.[3] He followed the projects of Thomas Edison in *Scientific American*, and in his first year of college, he looked to develop a better material for the incandescent lamp. In 1880, Edison had settled on carbon fiber for the incandescent lamp but was still continuing to search for a better material. Hall had correctly believed that tungsten might be that better filament. He would continue his work with tungsten, coming close to success, but found the search for commercial aluminum more promising. During this time, he had a summer job in Findlay and Bloomville, Ohio, selling a type of encyclopedia. Away from his chemicals, he still searched for an invention to make him famous. He developed an improvement on farm windmills, where a funnel of tin plate was put in front of the windmill to concentrate the wind. According to Hall's sister, "This was a valuable invention but Charles found that it had been already patented."[4]

Hall's chemistry professor, Frank Fanning Jewett (1844–1926), who gave Hall laboratory space to work on experiments, would prove to be an important mentor in his quest for commercial aluminum. The professor was a world traveler and as well educated in chemistry as any American academic of his day. Jewett, one of America's best chemists, had received his undergraduate education in chemistry and mineralogy at Yale University. From 1873 to 1875, he continued his chemistry studies at the University of Göttingen in Germany. It was there that Jewett met Professor Friedrich Wöhler, who had isolated aluminum as a metal in 1827. At Göttingen, he obtained a sample of aluminum metal. Upon returning from Europe, Jewett became Oliver Wolcott Gibbs's (1822–1902) private assistant at Harvard University. Gibbs was a famous American chemist known for performing the first electrochemical analyses, namely the reductions of copper and nickel ions to their respective metals. Gibbs was one of America's best mineralogists and chemists of the day. In 1876, Jewett was nominated by the president of Yale to teach science at the Imperial University in Tokyo. In 1880 at the age of 36, Jewett became the chair and professor in chemistry and mineralogy at Oberlin College. With Jewett in the chemistry department and Gray in the physics department, Oberlin would rival Yale and Harvard in science.

Frank Fanning Jewett and his wife, Sarah Frances Gulick Jewett, were one of the intellectual power couples of Oberlin's first century. Trailblazers and pioneers in their respective professions and actively involved in the community, they maintained a home that was characterized as the social and intellectual center of Oberlin. The Jewetts were very close to their students and rented out the attic and second floor of their home to them. Their home would also be the site of many discussions between Jewett and his student, Charles Hall. Jewett was an inspiration to Hall, and that role would be the key to the story of aluminum. Jewett had been studying aluminum since he first saw an engineer's transit made of it and the cap of the Washington Memorial at the Centennial Exposition of 1876. In Europe, Jewett had studied the efforts of chemists in preparing aluminum, including the possible use of cryolite as a flux. Jewett brought a pipeline of current technical information to Oberlin as well as a small button of aluminum from Deville's lab in France.

In 1920, Professor Jewett recalled meeting Hall that first summer prior to Hall's entering college, saying, "My great discovery has been the discovery of a man. When I went to Oberlin in 1880, on my return from four years teaching in Japan, there was a little boy about 14 years old [he was actually 16 at the time] who used to come to the chemical laboratory frequently to buy a few cents' worth of glass tubing."[5] At the time, glassblowing and glassworking was a necessary skill set for any young chemist, as very little commercial lab equipment was available. Hall was a typical Victorian scientist, using endless experimentation and observation to make steady progress. Like other Victorian scientists, he studied the published works of others, often taking experimental sidetracks to improve his overall understanding of science. Victorian scientists accepted little as proven theory until it was personally observed by experimentation. This almost obsessive experimentation often brought a deep knowledge but at a painfully slow pace.

Charles Hall had started studying the idea of electroplating, in which a current is passed through a metal salt electrolyte to deposit a metal on an electrode. It was chemically the reverse of the liquid lead batteries of his time. Silver- and gold-plating were commonly being used in the 1870s. The originators of silver-plating were George Richards Elkington and Henry Elkington, who began their research in the Industrial Revolution. By the 1830s, they had patented their processes; and in 1840, the technique of electroplating was brought to perfection. Electroplating silver was done by dissolving silver cyanide salts in water and then passing an electric current through it to have the silver ions (in solution) deposit on an electrode. Hall started with some simple experiments in electroplating to advance his knowledge of chemistry. For example, Hall built his own batteries, "out of all sorts of cups, tumblers."[6] When possible, he used ice cream ceramic crocks, which proved to be more durable. These were acid-lead batteries, requiring rebuilding of lead anodes daily. Unknown to him, his rival French competitor, Paul Louis Toussaint Heroult, was better funded and used the steam- and water-powered dynamos to generate direct current.

At the very end of his college career, Hall turned to aluminum as his best hope for fortune and fame. He had been playing with the idea for several years at the urgings of Professor Jewett. He had hesitated because experiments with aluminum would require molten electrolytes rather than plain water like he used in his electroplating work. He needed something more than the Bunsen burners available in Oberlin's chemistry labs. He would also need a surplus of higher-quality batteries or an electrical dynamo if he could get the money. There were other needs, too, such as bigger and better crucibles. Hall was aware that well-financed companies, such as the Cowles Brothers in Cleveland, had advanced electric furnaces to do the experimentation on aluminum. Hall's main advantage was his knowledge of chemistry, which in the end made up for his lack of electrical generating equipment.

Hall persisted with his experiments and studies throughout his undergrad work at Oberlin in various areas of science. He was often slowed by the need to build his own batteries to supply the current for his experiments and by the shortage of lab equipment. Hall was still searching for a prime project for extended study. He graduated in 1885, but he continued his research with the aid of his sister Julia. His mother died the same year he graduated, putting an additional strain on Charles and Julia to help with the family. It was at this point Charles decided to focus on his aluminum work, which offered the best hope for a breakthrough discovery. He had made substantial progress in his study of aluminum, and Charles now believed that the answer was using a thermochemical process with electrolysis. Hall recalled in his own words, "I had studied something of thermos-chemistry, and gradually the idea formed itself in my mind that if I could get a solution of alumina in something that contained no water, and in a solvent which was chemically more stable than the alumina, this would probably give a bath from which aluminum could be obtained by electrolysis."[7] He moved his home laboratory to a woodshed in the backyard.

The shed allowed him to install an old gasoline heater to heat crucibles of the alumina mixture.

Hall came to be an adept inventor from his interactions with Elisha Gray at Oberlin and his training in Victorian science. Hall documented the details of all his experiments, which was fundamental to the Edison approach to science and engineering of the period. For Hall, this methodology would eventually allow him to lay claim to the invention of the commercial process to produce aluminum. His approach was also like Edison's, with endless variations of experiments. Later in life, Hall would be faulted for his endless experiments following scientific curiosities, instead of practical research. He tried many fluorides before settling on cryolite, but he had Professor Jewett to discuss cryolite use with. He struggled with limitations of current generated by homemade batteries. His final improvement came with the use of a carbon crucible versus fireclay, which interfered with the chemical interactions. In a way, Hall's lack of standard equipment was a positive part of experimenting because it made him more flexible.

Paul Heroult, 1892 (courtesy of Carnegie Library of Pittsburgh).

In the mode of Victorian inventors, such as his idol Thomas Edison and his mentor Elisha Gray, Charles Hall became a passionate note-taker. He painstakingly detailed very experiment and often had his sister witness the recorded notes. He started the next experiment after a reading of earlier notes. He often went off on sidetracks to assure a theory or a principle was correct. Early biographer Junius Edwards writes: "Charles had developed the practice of recording in a notebook, various bits of chemical lore he noted in reading, but, more importantly, ideas of an inventive nature which flashed through his fertile mind. These ideas might be aids to a solution of an old problem with which he had been struggling for months or years, or might suggest new projects which looked profitable. Even in the midst of an absorbing interest in aluminum, Charles's mind kept turning to other subjects. Invention was at times a hobby that brought relaxation."[8] Often Charles used cryptic entries or notes for some of his best ideas. (He also had journal entries about life in general. Hall was so God-fearing that in his letters and journal entries, he referred to the devil as the "d-l.") This practice of detailed note-taking would be of importance in future patent battles.

On George Washington's birthday, February 22, 1886, Hall was able to produce a small amount of aluminum. The next day, he repeated the experiment, and he had his sister Julia witness his notebook. He took it a step further by writing a letter detailing all to his brother George in Dover, New Hampshire. The letter was signed, witnessed by Julia, and then mailed.[9] Later, this documentation would play a key role in winning a patent battle over his French competitor, Paul Heroult. Still, the process was far from commercially viable, and Charles lacked the money to further develop it or even start the patent process. Hall's original setup was only capable of a few ounces a day. Commercial production would require new anode material for electrolysis, electrical heating, a more powerful source of current, and higher-purity alumina.

Paul Heroult would arrive at the same point around the same time as Hall. Heroult was the same age as Hall, but had a number of advantages in his research. Heroult's father had managed a very successful tannery on the banks of the Orne River in Saint-Benin in France. Earlier in his life, Heroult's father had worked in an aluminum bronze plant that used the Deville process. Heroult studied chemistry at Saint-Barbe College, the same place that Deville had attended, and it was there that his interest in aluminum was sparked. Heroult was a far different man from Hall, but he shared the capitalist drive for fame and money via invention and discovery.

Christian Bickert of the French company Pechiney Aluminum described Heroult as having "none of the attributes of the traditional scholar. He was high strung, unruly, occasionally hard and insolent; he did not fit the image of wise, disciplined men of science. He loved games, the company of women, travels by land and sea; he was a free spirit in an impetuous body. No comparison with the austere scientist struggling with stubborn mysteries. His discoveries were not the result of long sleepless nights spent in the laboratory, or of complicated demonstrations. Heroult loved life, and could not have borne such restrictions. Instead, his inventions appeared suddenly, out of the blue, a stroke of common sense, or of genius, sometimes during a lively game of billiards, his favorite pastime."[10] Clearly he was not out of the mold of Victorian scientists such as Thomas Edison, but more like Edwardian high-living inventors such as Nicola Tesla. Nevertheless, Heroult was an engineer at heart. He was inspired by a famous work, *Sainte-Claire Deville's Aluminum, its properties, its production and applications.* For Heroult, money was a far better motivator than fame.

Heroult was a student at the École des Mines, Paris, when he began working on the electrolysis of aluminum compounds. Heroult quickly came to the conclusion that aluminum was the path to riches at the expense of the other subjects he should have been studying. As a result, he failed his first year and was dismissed from school. When his father died unexpectedly during 1883, leaving him the tannery buildings, he decided to use them as a laboratory for his experiments. He induced two fellow students from l'Ecole des Mines, Louis Merle and Lucien Van Kerguistel, to join him, and they began to work on electrolysis.

His mother gave him 50,000 francs to acquire the Bréguet dynamo he needed. The dynamo supplied a consistent and powerful source of electrical current for his experiments, which gave him a huge edge over Hall, who spent hours building batteries. However, he struggled with all the chemical problems that confronted Hall. Aluminum was produced sporadically until Héroult found a molten bath using cryolite created the right environment. The aluminum oxide was then reduced, fell to the bottom of the crucible, and triggered the joining of the new metal into a mass. With further work in 1886, he discovered that electrolysis of a solution of alumina (aluminum oxide) in molten cryolite (sodium aluminum fluoride) resulted in the formation of a layer of molten aluminum at the bottom of the vessel. Electrolytic aluminum was born. And without missing a beat, or fully proving his discovery, Héroult applied for a patent.

Clearly, Heroult did have better equipment, like a state-of-the-art dynamo, to generate his electricity. The dynamo generated an electrical current by moving a magnet back and forth in a copper wire coil. This allowed a dynamo to generate steady, usable direct current. The best dynamos were driven by steam engines, as in Heroult's case. The tannery had a steam engine available to power the dynamo, whereas Hall was limited to low-power, homemade batteries. In the 1870s, dynamos had become commercially available. The dynamo would prove to be a key step in the discovery and commercialization of aluminum. While Charles Brush used his DC dynamos and arc lights to light a park in Cleveland, Ohio, and also built the dynamos used by the Cowles Brothers to produce aluminum bronze, Albert Schmidt was a Swiss mechanical engineer who developed the DC dynamos in Europe. Schmidt had worked on Europe's earliest dynamos and arc lighting systems in the 1880s. Eventually, Hall would also have a dynamo to support larger commercial experiments at the Cowles Brothers plant in Cleveland.

Heroult's approach to experimentation allowed him to move faster towards the goal. Heroult was more engineer than scientist, the opposite of Hall.

Heroult moved forward on his quest with much less experimentation, using the published scientific papers to plan advanced experiments. He had little interest in the methodology of endless trial-and-error experimentation and documentation. Here again, we see the similarities of Heroult to men like Tesla and George Westinghouse versus Victorian scientists like Edison. Heroult had clearly evolved past the trial-and-error methods of many early Victorian inventors as he started to use science to narrow the scope of experiments needed. This is another example of pioneering in the methodology of modern research and development. Men like Edison spent endless hours in trial-and-error experiments, while Heroult eliminated many trials by the application of science. Hall, however, had developed a body of knowledge, which would help commercialization move forward.

Heroult, Tesla, and Westinghouse were the philosophical leaders of a group of inventors who defined the new field of engineering. The generations before them were true scientists, while they looked for applications before theory was fully understood. Their skills included creativity in applying scientific principles and theories. The individuals in the Heroult group were not specialists, but creators willing to chase economic needs and practical challenges in any scientific discipline, and they believed that inventing was a profession itself. Invention was seen as a craft, which would become the discipline of engineering. Heroult and Westinghouse, more than any of the great Victorian inventors, pioneered the discipline of the engineering craft. Their approach would evolve into the corporate approach to research and development used today.

Captain Alfred Hunt, 1890 (courtesy of Carnegie Library of Pittsburgh).

Amazingly, Heroult, the engineer, and Hall, the scientist, would arrive at the same point at the same time. Heroult would apply for his patent in April of 1886, ahead of Hall's filing. Hall's records and documentation, however, proved he had discovered the process two months earlier. Heroult's process was a bit different, using carbon anodes and internal electrical heating, but like Hall's, it was far from commercialization. Heroult proved less patient and moved on to other projects; however, Heroult's efforts to find investors failed. A.R. Pechiney of Merle Chemical Aluminum Company told Heroult that "the market for aluminum was limited to opera glasses," and he should focus on aluminum bronze.[11] Hall stubbornly held to his quest, though, while Heroult wanted to sell his patent quickly and take the profits. As in America, there were a number of French companies making aluminum bronze (mostly composed of copper).

Merle Chemical was making aluminum bronze as well as some aluminum utilizing the Deville process that Napoleon III had financed. Héroult offered to sell the process to A.R. Pechiney, but he declined to buy it. Héroult subsequently sold his patent to another company, the Société Électrométallurgique Française, which built its first aluminum factory in Froges, France. In 1889, faced with competition from Froges, A.R. Pechiney closed down Merle's Chemical Aluminum Department. Then, in 1897, Pechiney bought Société Électrométallurgique Française and entered the field of aluminum electrolysis, using the original Heroult process. Charles Hall's path to commercialization would be far better.

Nonetheless, Charles Hall had also hit a road block. He had discovered a cheaper method, but he was in no position to go into production. For both men, commercialization of their discovery would require capital and even fine tuning of the process for mass manufacturing. Charles Hall had the patent, but he realized that the real prize was commercialization. He knew well from Elisha Gray that the real prize and glory would go to the first person to commercialize the process. Without the ability to commercialize his process, Hall would be but a footnote in the history books.

Chapter 4

Commercialization

THE COMMERCIALIZATION OF the Hall-Heroult process is often overlooked in general history books. Historian and aluminum pioneer Alfred Cowles stated: "The generally accepted statement that Hall, in his woodshed experiment of February 23, 1886, discovered the modern process for producing aluminum is, to say the least, an oversimplification."[1] Commercialization required investors, capital, and more development work. It would be slowed and plagued with endless years of ligation between Hall and Heroult, as well as others like the Cowles Brothers. The futuristic predictions of aluminum applications remained far from reality in the 1880s. New process technologies, such as aluminum rolling and extrusion, would be needed to make these applications reality. Even the Hall-Heroult process was far from producing much more than a handful of aluminum in a day. Where Heroult lacked the determination and patience, Hall seemed fully up to the task. Hall's first stop would be the Cowles Brothers, who dominated in aluminum bronze production and electric furnace applications in the development of alloys. In 1884, the brothers developed one of the world's first metallurgical electric furnaces. They were also world experts on the use of electric furnaces and the production of aluminum bronze. Additionally, the Cowles Brothers had the most powerful electric furnaces in the world.

The Cowles Brothers had been working with Brush Electric in electric power generation in an effort to make more powerful furnaces. They teamed up with Brush Electric to reduce zinc ore into zinc metal using the hydroelectric power of the Pecos River. Within months, they were working with aluminum ores to produce aluminum bronze. On March 18, 1885, the brothers organized the Cowles Electric Smelting and Aluminum Company and took out patents for the electric furnace processing of aluminum bronze. In 1885, the Cowles Brothers opened an aluminum bronze plant in Lockport, New York. The site was 15 miles from Niagara Falls, and tailrace of the falls supplied cheap water power for their huge Brush dynamos. These new and powerful furnaces were being used to experiment with pure aluminum production. At the time, Hall was unaware of the Cowles Brothers' work on pure aluminum.

The Cowles's dynamos were the world's largest and allowed the brothers to produce an amazing ton and a half of aluminum bronze per day. The plant cut the price of aluminum bronze to 40 cents a pound, about a third of the market price in 1885. In January of 1886, the Cowles Brothers predicted prematurely that they would soon be producing pure aluminum at 60 cents a pound versus the 1886 market price of $9 a pound. In 1887, Hall, having failed to develop a commercial process for his first investors, turned to a working arrangement to experiment at the Cowles' powerful Lockport electric furnaces. At the time,

Founders of Pittsburgh Testing: George Clapp (left) and Alfred Hunt, 1893 (courtesy of Carnegie Library of Pittsburgh).

the patent office had also notified Hall of "interference" with the Heroult patent, and the Cowles Brothers offered their lawyers to help Hall in the legal struggle.

The deal with the Cowles Brothers included a salary of $75 a month. The arrangement also included an option for the Cowles Brothers to obtain the patent rights in return for Hall's having an eighth share in the formation of an aluminum company if successful. From the start, there was a lot of distrust between the two parties, but Hall and the Cowles Brothers learned from each other. Moreover, though, the patent battle affected their relationship. Should Heroult win, the Cowles Brothers would have no need for Hall and his process.

The arrangement with the Cowles Brothers was marked by technical differences in the commercial development of aluminum as well. Hall used copper anodes, as opposed to the carbon ones preferred by the Cowles Brothers (as well as Heroult). Furthermore, Hall refused to utilize the full efficiency of the electric furnace by using it to melt and heat the ore mix; instead, he used external heating to melt his ore mix. Hall experimented at the Lockport operation for 12 months without much success[2] and, in a year, he failed to produce commercial amounts of aluminum. The relationship never amounted to anything except another series of ligations between the Cowles Brothers and Hall. Hall did become close friends with the plant manager, Romaine Cole, who would eventually be an imperative key to commercialization.

Romaine C. Cole had experimented in the 1880s with aluminum through the use of an open-hearth furnace in an unsuccessful attempt to convert alumina (that is, aluminum oxide) into aluminum. Cole had been experimenting at Pittsburgh Testing Company under the direction of the prominent metallurgist Alfred Hunt (1855–1899). It would be Cole who would bring Hall and Hunt together. These three men would be the seed of the great Aluminum Company of America (ALCOA), and their meeting took place on July 27, 1888.

Alfred Hunt, aka Captain Hunt, based upon his service in the Pennsylvania National Guard, was a New Englander by birth and graduated from the Massachusetts Institute of Technology in 1876 with a degree in metallurgy and mining. He started with Bay State Ironworks, which was operating the first open-hearth steel furnace in the United States of America. From there, he would go on to Nashua, New Hampshire, to work for the Nashua Iron & Steel Company. He gained a reputation early as an expert metallurgist. He eventually came to Black Diamond Steel in Pittsburgh at the request of a friend of the family, Pittsburgher James Park, Jr. Black Diamond Steel was owned by Park, Brother & Co. The Park family was part of Pittsburgh's old Pig Iron Aristocracy since 1804 and had deep ties with Pittsburgh's East End capitalists. The original Park factory had been one of the first to make the transition from a blacksmith shop into an iron manufacturing plant in Pittsburgh.

The Park family had been the first investors in Pittsburgh baseball, the Allegheny Observatory, and the University of Pittsburgh. James Park, Jr., was the founder of the Chemistry and Mineralogy Department at the University of Pittsburgh in the 1860s. Later, James Park's fight to support the patent battle of Pittsburgher William Kelly over Bessemer would result in a victory in 1898, although Kelly would be lost to history. Hunt's relationship with James Park opened many doors for the young man. He would be accepted into the core of Pittsburgh's aristocracy, which included the Masons, Presbyterianism, the pig iron industry and large sums of Pittsburgh's East End capital.

Alfred Hunt home at 272 Shady Lane, Pittsburgh, where ALCOA was founded (courtesy of Carnegie Library of Pittsburgh).

In 1886, Hunt did some experiments on making aluminum in an open-hearth furnace. When Hunt joined Black Diamond Crucible Steel, it was making specialty steel that was considered on a par with that of Sheffield, England.

Hunt rented an apartment in the upscale boarding home of the Negley family in Pittsburgh's East End (the home of many of America's greatest industrialists of the time). He joined the Presbyterian Church and the Masonic Lodge as he rose through Black Diamond. He became friends with another metallurgist and chemist at Black Diamond named George Clapp (1858–1949). At the time, Clapp's father was the company's treasurer. One of the jobs being handled by the Black Diamond was that of producing the highest quality of steel required for a series of new bridges that were being erected across Pittsburgh's rivers by the famous builder, Dr. Gustave Lindenthall.

George Clapp was the son of DeWitt Clinton Clapp, an established iron and steel man and executive. He graduated from the Western University of Pennsylvania (today's University of Pittsburgh) in 1877 with a degree in chemistry. Years later, Clapp would be instrumental in the offering of a degree in metallurgical engineering. Clapp already had strong ties to Pittsburgh's industrial wealth, and his interests were more in business than metallurgy. Clapp would become a good friend and partner with Alfred Hunt, and Clapp was able to introduce Hunt to the old Pittsburgh aristocracy.

Hunt and Clapp had positioned themselves as young and upcoming Pittsburgh East Enders. The East End of Pittsburgh was known as the world's richest neighborhood, and included amongst the neighbors were Andrew Carnegie, the Mellon family, George Westinghouse, and H.J. Heinz. Other neighbors included: Henry Clay Frick, founder of United States Steel; George Mesta, founder of Mesta Machine Company; Jacob Vandergrift, partner of Rockefeller in Standard Oil; Charles Lockhart, co-founder of Standard Oil with Rockefeller; Sylvester Marvin, founder of National Biscuit Company (Nabisco); James McCrea, president of the Pennsylvania Railroad; Robert Pitcairn, vice-president of Pennsylvania Railroad and industrial investor; John Pitcairn, president of Pittsburgh Plate Glass; Oswald Werner, developer of dry-cleaning; William Lash, president of Carbon Steel and co-founder of ALCOA; Willis McCook, co-founder of Pittsburgh Steel; Wallace Rowe, co-founder of Pittsburgh steel; Alexander King, glass magnate; David Stewart, president of Pittsburgh Locomotive; Henry Hillard, president of Alcania; Henry Laughlin, Jones and Laughlin Steel; Alexander Bradley, stove manufacturer; Benjamin Thaw, railroad magnate; James Guffey, oil and gas magnate and co-founder of Gulf Oil; Julian Kennedy, world blast furnace expert; George Macbeth, founder of Macbeth-Evans Glass; Daniel Clemson, a Carnegie partner; Benjamin and Thomas Bakewell, Pittsburgh's famous glassmaking family; Thomas Armstrong, the founder of Armstrong Cork; Thomas Messler, president of the New York and Erie Railroad; Lillian Russell, national theater star; and Joseph Woodwell, hardware baron and artist. Others included Durbin Horne, president of Joseph Horne Department Store; Alexander Moore, newspaper baron (as well as ambassador to Spain); James McClelland, famous homeopathic physician; Arthur Braun, publisher and banker; Thomas Howe, president of the Exchange National Bank of Pittsburgh; George Berry, president of Citizen's National Bank; Rueben Miller, president of Fidelity Trust Company; August Succop, banking executive; James Callery, president of Pittsburgh Railways; and bank president John Holmes. Many of these neighbors would become investors in Hunt and Clapp's future company.

Hunt and Clapp formed a new company doing metallurgical work, Pittsburgh Testing Laboratory, and they would acquire a partnership in 1887. Pittsburgh Testing was a consulting and testing firm organized to service the steel and railroad industry of Pittsburgh. Clapp was particularly adept at getting capital from the company. Hunt was focusing on the development of high-quality steel and the use of electric furnaces in steelmaking. Hunt had studied the improved quality of steel by using deoxidizing agents such as silicon. He was aware that aluminum was a more powerful deoxidizer, but there was no commercial supply, and that would lead him to study aluminum further as a metal.

Pittsburgh Reduction Company, Pittsburgh's north side, 1891 (courtesy of Carnegie Library of Pittsburgh).

Hunt's interest in aluminum was twofold. First, like Hall, he saw a commercial value in the metal itself, and second, he saw a potential to use aluminum in the manufacturing of high-quality steel. Its use in the deoxidation of steel offered an immediate market for pure aluminum. This would become a hidden factor in the commercialization of aluminum, one that Alfred Hunt, George Clapp, and their associates were uniquely positioned to exploit with their steel and railroad contacts. Hunt also had his old ties with the Cowles Brothers' work on aluminum with Romaine Cole. The early work of Romaine Cole and Hunt had focused on aluminum to use in steelmaking as well as the use of electric furnaces for steelmaking. Hunt had tested samples of aluminum deoxidized and found an amazing level of quality and a new level of ductility in the forming of steel. This aluminum deoxidized steel rivaled the quality of the great Krupp Works in Germany.

Hunt realized that pure aluminum could be used for deoxidation of the liquid steel, if a pure supply of aluminum could be found. He had been working on the idea at Black Diamond Steel, but there were limited amounts of aluminum even for experiments. With limited quantities, Hunt had proven aluminum's use in making high-quality steel. The secret was, aluminum helps in removing the dissolved oxygen from the liquid steel, a process known as killing. Silicon was also used for the similar purpose, but Hunt had shown aluminum was more efficient and produced a much higher quality of steel. While silicon took oxygen out of steel to allow easy casting, it produced large, dirty inclusions in steel in the process. Aluminum, combined with oxygen bubbles, prevented the formation of inclusions and defects during the casting of steel. The elimination of inclusions greatly improved the mechanical properties of steel. And, ironically, the use of aluminum in steel would help make it more competitive with aluminum.

The problem for Hunt was the high price of aluminum and its lack of availability. In the 1880s, at Diamond Steel, Hunt and Romaine Cole had made unsuccessful attempts to convert alumina (aluminum oxide) into aluminum through the use of an open-hearth furnace. However, Hunt had never fully given up on somehow producing aluminum. Romaine Cole would introduce Charles Hall to Alfred Hunt, who could offer a solution.

Romaine Cole and Alfred Hunt would meet with Pittsburgh industrialists in the East End. This small group met on July 31, 1888, in Captain Hunt's living room at his home in Pittsburgh, and decided to go into the business of making aluminum by the Hall process. Romaine Cole would bring the big investors to the table, as well as his own capital. These investors were Howard Lash (president of Carbon Steel), Millard Hunsiker (sales manager of Carbon Steel), Robert John Scott (a Carnegie partner and superintendent of Union Mills), and W.S. Sample (chief chemist at Pittsburgh Testing). The meeting was held at Hunt's home located at 272 Shady Lane (Shady Avenue) in East Liberty (Pittsburgh's Eastside). This organizational meeting resulted in the formation of Pittsburgh Reduction Company (future ALCOA), whose name would be known for the commercialization of the Hall process. The six partners paid a total of $20,000, and the 10,000 shares were initially divided up as follows: Hall, 3,525 shares; Cole, 1,000 shares; Hunt, Clapp, Hunsiker, Lash, Sample and Scott had 3,006 shares; and the rest went to a number of rich East Enders. Within two years, additional stock was issued to over 30 more East End investors. Initially, Hunt was named president of the company, and although Hall was not an officer, he was in charge of plant operations and, of course, the largest shareholder.

Years later, the Pittsburgh connection in supplying the people and capital would be hailed: "Aluminum is a Pittsburgh product—not because Pittsburgh had abundant supplies of bauxite ore, for she has not; and not because Pittsburgh had abundant hydroelectric power, for she has not. Aluminum is a Pittsburgh product because Pittsburgh, despite its reputation for smoke and grime, is primarily interested in men. Here in Pittsburgh as in no other community of the United States, does creative genius get a hearing and sound backing.... It was Pittsburgh that listened to an Ohio college boy with the vision of the possibilities of a new and light metal. Not only did it listen, but it gave substantial assistance."[3] Not surprisingly, two other major companies, Standard Oil and Carnegie Steel, were born in the same wealthy neighborhood.

It had taken a few years, but Hall had finally found capital for his vision. In addition to being the major stockholder, Hall was also named plant manager at a salary of $5 a week. A few months later, a friend of Hunt, Arthur Davis (1867–1962), was brought in to help Hall in the daily management of the operation, which allowed Hall to focus and work more on the technical issues. Davis proved to be the cornerstone of the company and soon became general manager of the firm, and was named director in 1892. He worked as general manager when the firm became the Aluminum Company of America (ALCOA) in 1907; he became president in 1910 and chairman of the board in 1928, a capacity he served in until 1958.

Originally, Charles Hall rented a room over a saloon for $5 a month, but Alfred Hunt thought Hall should live in a better neighborhood. Charles Hall took up residence in an East End apartment house owned by Hunt. A few months later, two other old-time East Enders, Andrew Mellon (1855–1937; banker) and William Thaw (steel baron), would invest capital. Captain Hunt had come to Andrew Mellon for a small loan to help with the operation. Andrew Mellon, like so many of the Gilded Age, was captivated by this silver-like metal. He offered a much larger loan and later bought $6,000 worth of stock. It was the beginning of an historic relationship.

The new company started operations on Smallman Street (on Pittsburgh's Northside). Yet another Pittsburgh East Ender, George Westinghouse, supplied the two powerful Westinghouse dynamos to produce the needed electrical current. The 1,200 amperes and 25 volts of each stream-driven dynamo was beyond any laboratory battery and offered the potential for a continuous operation. George Westinghouse himself supervised the installation. Westinghouse had been experimenting with aluminum parts for several years and looked forward to the availability of aluminum. Davis and Hall would develop a lifelong friendship. On Thanksgiving Day 1888, the plant, under the supervision of Charles Martin Hall and Arthur Davis, produced its first commercial batch of aluminum. Hall's long hours would end his on-and-off engagement to his college sweetheart, Josephine Cody. Hall was what we call today a workaholic, and would never marry. Instead, he would be forever married to his laboratory with very few other interests.

In early 1889, the company was struggling to make 30 pounds a day at a selling price of $8 per pound. Production was kept in the company's large safe. It was hard to keep workers due to the fumes, which proved to be more toxic than those of the neighboring Carnegie steel mills. The production was far

Original dynamo for Pittsburgh Reduction Company, 1893 (Library of Congress).

below the original predictions of 250 pounds a day. By mid–1889, the Smallman plant was producing about 1,000 pounds of aluminum a month at approximately $5 a pound. Most of the raw ingot aluminum was being sold to steelmakers, and its first major customers were Carbon Steel Company and other Pittsburgh steel companies. By the end of the year, the price was around $2 a pound, and Hall was able to reach 50 pounds a day. The company reported a loss, but still ordered larger Westinghouse dynamos to improve production. In August of 1890, the company reported a loss of $637 from the prior year.

The newest problem to arise was the lack of aluminum finished product demand. Before the commercial breakthrough of aluminum, a young Alfred Hunt had published a paper predicting aluminum buildings, airplanes, and pots. Hunt even believed that aluminum horseshoes might give his racehorses a competitive advantage.

Interestingly, one of the first commercial users of aluminum would be by another East End neighbor of Hall and Hunt, George Westinghouse. George Westinghouse was extremely excited about the new availability of aluminum that his AC electrical system had created. Westinghouse anticipated that cheap power would mean low-cost aluminum, and he and his engineers were looking at potential uses. Aluminum additions to Babbitt bearing metal showed superior results at Westinghouse Machine and Air Brake Companies. These engineers were also using aluminum bronze in corrosion applications. Westinghouse soon became a customer of Pittsburgh Reduction experimenting with aluminum parts. One of his first applications was a valve in his air brake, and he also found it to be useful as a shim for machinery. Westinghouse's foundry even looked to it as a casting metal to replace cast iron.

Another American application that Pittsburgh Reduction Company was the first to use was aluminum foil. Alfred Hunt had envisioned possible aluminum foil uses in the 1880s. George Westinghouse had studied the production of aluminum foil in France and Germany and had suggested its manufacture to Pittsburgh Reduction. Westinghouse, and a number of his Pittsburgh East End neighbors like Henry Clay Frick, used aluminum foil on walls when it had a price higher than gold. The aluminum wall decoration of that period can still be viewed today at the Frick mansion in Pittsburgh. Still, Pittsburgh Reduction was slow to advance their rolling technology for foil, and the Europeans were already the leaders in tin foil production in the late 1890s.

Foil made from a thin leaf of rolled tin was commercially available before aluminum. Tin foil was marketed commercially in the late 19th century. It had been around for decades as Pittsburgh Reduction started to make it in limited quantities. The term "tin foil" survives in the English language as a term for the newer aluminum foil. Tin foil is less malleable and harder than aluminum foil, and tends to give a slight tin taste to food wrapped in it. The first aluminum foil rolling plant was opened in Emmishofen, Switzerland, in 1910 as part of an earlier aluminum plant. The plant, owned by J.G. Neher & Sons, the aluminum manufacturers, started in 1886 in Schaffhausen, Switzerland, at the foot of the Rhine Falls, capturing the falls' energy to produce aluminum. The earliest use of aluminum foil was leg banding and message holders in racing and homing pigeons. The Germans were the first to use it in packaging of chocolate bars. In 1898, Pittsburgh Reduction held off on the further development of aluminum foil to use their production capacity for castings and aluminum's use in steel production.

Alfred Hunt would leave for a year in 1898, as the sinking of the battleship USS *Maine* in Havana took the United States into the Spanish-American War. Hunt had training in artillery at MIT, and he would volunteer and command an artillery unit. Aluminum would make its first war appearance in the Spanish-American War (1899) thanks to Hunt. Teddy Roosevelt charged up San Juan Hill with an aluminum canteen, and the army was using aluminum tent pins at the time. It was a short war, and after a year, Hunt returned, marching in victory down Pittsburgh's Fifth Avenue. Hunt's return sparked new energy into the operation. Soon after, Hunt traveled to Europe to sell Hall's European patent rights for the cash that supplied the money to continue work and expand the Pittsburgh operation. Hunt would steer Pittsburgh Reduction Company forward, but his tenure proved short: he died one year later from complications from the malaria he had contracted during the war.

Hunt was more than a founder and company manager. He was one of America's greatest metallurgists who took aluminum to so many different applications. He was a prolific writer on engineering and metals and spoke at conferences around the world. Additionally, he foresaw almost every use of aluminum we see today. In the first year, Hall and Hunt literally lived at the Smallman plant to make it a success. Over the years, Hunt became close friends with Charles Hall. While Hall tended to prefer the laboratory, Hunt was comfortable in any phase of aluminum production. Hunt had a command of marketing, operations, and development. He was a natural leader, which allowed him to build a powerful company. He had built important alliances with the Mellon family, and the Mellons trusted him in making all operational decisions.

The loss of Hunt as company president was a blow to the Mellon family, which had about 30 percent of the stock. The Mellons saw Charles Hall as a scientist, not a manager, and Hall would have agreed. A.V. Davis was considered too young to be president when Hunt died in 1899, so Richard B. Mellon became president. Davis, however, served as the general manager of operations and would become president in 1910.

The growth of the company after the Spanish-American War was phenomenal. A larger production plant was soon needed. The decision of locating at New Kensington on the Allegheny River was heavily influenced by Andrew Mellon and his brother Richard. The Mellon Bank's namesake family were already one of America's wealthiest. Thomas Mellon, the founder of Mellon Bank, had made fortunes in coal, oil, iron, and steel. Thomas Mellon's son, Andrew Mellon, would rival J.P. Morgan in money and power. The Mellons controlled most of Pittsburgh's industries as well as the city itself. Thomas Mellon was said to use a military style in preparing his sons for business, and allegedly, the Mellons only broke dinner table silence to speak business. The social bonds seemed to be more Scotch-Irish ancestry, Presbyterianism, and Freemasonry, than financial. Andrew Mellon's best friend was Henry Clay Frick, who was Carnegie's partner. Andrew Mellon and Frick partnered in many real estate deals, becoming major landowners, and eventually formed an investment company. In the 1890s, Andrew Mellon was looking to make his own mark in business, and he would do so within the aluminum industry. Andrew Mellon quickly seized the idea and opportunity of combining real estate and aluminum. T. Mellon & Sons owned large amounts of land on the Allegheny and were investing in land development companies. The Mellons offered the aluminum plant four acres of level ground on the river and a $10,000 cash bonus to locate there.

The original Smallman plant had struggled to produce a few cooking utensils, but its first product was a teakettle. The molds were borrowed from the Griswold Company of Erie, Pennsylvania, a manufacturer of cast iron cookware, where some aluminum teakettles were also made. Griswold Company was impressed and placed an order for 2,000 kettles, but then pulled out of the deal. Griswold Company wanted the kettles not to be made of aluminum, so ALCOA went into the fabricating business to prove that there was a market for this metal. Arthur Davis replaced the loss of Captain Hunt in 1899 by aggressively expanding research and development. The New Kensington plant was specifically designed to manufacture sheet aluminum to make cooking utensils. By 1890, even with the prior losses, the Mellon family investments brought other wealthy investors in.

Andrew Mellon in 1927 (Library of Congress).

Pittsburgh Reduction spent a lot of effort trying to develop a market for aluminum-made cooking utensils. A cluster of companies had developed in Wisconsin, but others had failed in their labors to sell aluminum cooking utensils. In 1901, Pittsburgh Reduction Company organized the Aluminum Cooking Utensil Company to sell products to housewives. The company developed a novel plan to sell its Wear-Ever brand, and college students were recruited to sell door-to-door by demonstrating aluminum cookware. Door-to-door and department store demonstrations were required to let the public become comfortable with this light metal because it was a tough sell for the housewives who had grown up cooking in cast iron.

Alfred Hunt was an eccentric and passionate metallurgical engineer who built a foundation for aluminum products prior to his death. He had aluminum sinks and bathtubs installed at his East End home as he tried to market these new applications with his East End neighbors. Arthur Davis also promoted the local use of aluminum in his Pittsburgh home and would take Hunt's place as product champion. Andrew Mellon, a future major investor in the company, would also install aluminum bathtubs and sinks, in addition to being the owner of one of the first aluminum cars. The aluminum foil that Mellon saw in the homes of Henry Clay Frick, Westinghouse, and other industrialists had always fascinated him. Andrew Mellon had become a believer in aluminum and backed the company loans in very difficult times. Interestingly, Frick would turn down the aluminum investments that Mellon brought to their weekly Duquesne Club investment meetings. It would be one of the few times that Frick passed up a successful investment. By 1894, Mellon owned 12 percent of the company and controlled all the loans. Andrew Mellon would soon match Alfred Hunt in his passion for aluminum.

In later years, Andrew Mellon converted many decorations in his mansion to aluminum and ordered the first all-aluminum cars to be made for his managers. Andrew Mellon wanted to move into additional companies versus his father's banking relationships with industry. Andrew Mellon had achieved a foothold in a Pittsburgh company that his father had failed to do with Westinghouse Air Brake. Westinghouse would not have allowed a Pittsburgh banker on the board, nor would Carnegie allow Mellon on his board. Mellon, like J.P. Morgan, believed banking and business should form alliances, and he always looked for working partnerships. This model of banking and industry became known as the "Mellon System." Richard Mellon also became a major investor with his brother, Andrew, in 1892. Andrew Mellon went on to form Koppers, Carborundum, Union Steel Company, Gulf Oil, Standard Steel, Pittsburgh Coal, McClintic-Marshall, and a host of other sizable companies applying the successful model of management. The Mellon family's involvement and success in ALCOA assured a stream of investors.

In Europe, the aluminum industry would take a much different path. Charles Hall was not alone in finding investors like Andrew Mellon in aluminum;

Heroult also found some success in Europe. However, Heroult would never be a dominant investor himself. Heroult had to sell patent rights in France and start a Swiss joint venture to start the commercialization of his process. In 1889, Paul-Louis-Toussaint Heroult, Gustave Naville, Georg Neher, and Peter Emil Huber established a company called Aluminium Industrie Aktien Gesellschaft (AIAG), and later known as part of the Alusuisse-Lonza group in Zurich, Switzerland, to extract aluminum, creating the first aluminum production plant in Europe. It established plants in Neuhausen am Rheinfall in 1888, in Rheinfelden, Germany, in 1898, and in Lend, Austria, in 1899. In 1899, the company started to invest in the Valais region of Switzerland because it was rich in hydropower resources to power its aluminum plants. It would be a perfect match for the natural resources of Switzerland. Aluminium Industrie Aktien Gesellschaft (AIAG) became the root of the great Swiss aluminum company Alusuisse-Lonza Holding AG, merged into Alcan in 2000.

Another aluminum company that Heroult and his father worked for was Pechiney. The company was originally founded in 1855 by Henri Merle, a producer of caustic soda at a manufacturing facility in Salindres. First named as Compagnie des Produits Chimiques Henri Merle, the company was renamed in 1897 as the Société des Produits Chimiques d'Alais et de la Camargue. The company first began producing aluminum metal in 1860, using a chemically based process developed by Henri Sainte-Claire Deville in 1854, and was granted a 30-year monopoly by the French government. Pechiney, after initially turning down Heroult, ten years later, purchased Société Électrométallurgique Française, the company Heroult had sold his patent to. In 2003, the remnants of this company were merged into Alcan. The good news for Europe in 1899 was that there were three competing companies in three countries. However, the success of the Hall and Heroult processes would lead to a decade of legal battles by the turn of the century.

Chapter 5

Litigation and Growth

THE CONFLUENCE OF TECHNOLOGY that opened up commercial aluminum would, by its nature, create a number of legal claims. The legal struggle would consume decades, but the Mellon family would provide the legal backing as men like Hall, Hunt, and Davis moved the operations forward. On April 2, 1889, Hall and Pittsburgh Reduction won a major legal battle when the patent office ruled Hall had priority over Heroult on the American rights for the aluminum making process. This should have assured the company smooth sailing in America for aluminum production for 17 years; however, many more claimants, such as the Cowles Brothers, would appear. For Pittsburgh Reduction, it was not a matter of honor or glory, but survival and profitability; and therefore, legal compromises needed to end. Even with the legal uncertainties, Pittsburgh Reduction needed to improve on its Smallman Street plant to stay competitive with European aluminum, as Heroult's patents still had priority in Europe.

The Cowles Brothers would challenge the Hall patent by using the process and underselling Pittsburgh Reduction, forcing the company to file suit against the brothers. At the same time, Pittsburgh Reduction needed something better than its experimental plant on Smallman Street. Charles Hall argued at a stockholders meeting that the company had to move forward, even with the uncertainty of the legal issues. He reasoned that if the company lost its legal claim, aluminum making would be a "free art," and this would require efficiency to compete.[1] The New Kensington plant had to be built with its new cells for aluminum production to stay competitive. Drawn-out legal battles would be costly, but the decision was made to move forward on the New Kensington plant with the help of new investors such as Andrew Mellon and others.

The New Kensington project required both a factory and a new town. The city of New Kensington was a focused industrial town built and incorporated in 1892. Its location was 18 miles northeast of Pittsburgh. The development of New Kensington began in 1890 when the Burrell Improvement Company, an affiliate of Mellon Bank, purchased level land on the east side of the Allegheny River as a prime location for a city. They had the land surveyed and laid out the town of "Kensington." The town was located on the Allegheny River and had access to natural gas and coal fields to fuel industry. The land between Second Avenue and the river was to be maintained in larger pieces for sale to industrial users. While Pittsburgh Reduction would dominate, it was not to be the only industry in town.

The first public sale of lots took place on June 10, 1891. Purchasers were given a free train ride from Pittsburgh, and refreshments, if they came to view the site of the proposed new town. The price range of the first several hundred lots ranged from $30.00 to $300.00. By the end of 1891, New Kensington would be home to 12 companies that provided jobs to 4,000 individuals. In addition to the Pittsburgh Reduction Company (PRC), the other companies included the Bradley Stove Works, the Brownsville Plate Glass Company, Kensington Chilled Steel Company, Kensington Roller Process Flour Company, Kensington Tube Works, Logan and Sons Planning Mills, New York Piano and Organ Factory, Pennsylvania Tin Plate Company, the Rolled Wheel Steel Company, the R.F. Ryan Planning Mills, and the Chambers Glass Company.

The huge New Kensington plant was able to produce 1,000 pounds of aluminum each day by early 1893. It had established a continuous operation using the coal-fired steam plant to drive four Westinghouse generators to run 20 pots. The plant produced raw aluminum and had forging, rolling, tube, and casting departments to manufacture finished products. The plant also was equipped with a raw materials department to improve on the purity of the received aluminum ore. This vertical integration would become the hallmark of company strategy. The New Kensington plant gave Pittsburgh Reduction Company an edge in American competition, but there was still strong competition that threatened the young company. Much of this competition was coming from Europe with raw aluminum and American fabricating, rolling mills, and foundries. Europe had already gained an early cost advantage with cheap hydroelectric power, while ALCOA used costly coal and electric generators. In addition, the Cowles Brothers were supplying cheaper aluminum domestically in a price war environment. Pittsburgh Reduction was nearing collapse under the fierce competition. Andrew Mellon and rich Pittsburgh East End investors would supply the capital and political power to save Pittsburgh Reduction for the future.

Andrew Mellon and his Republican ties proved critical in getting aluminum added to the McKinley protective tariffs. Mellon was able to get 15 cents a pound tariff on aluminum during the critical years of the 1890s when cheap

European aluminum could have destroyed the infant company of Pittsburgh Reduction.[2] Even with this protection, Pittsburgh Reduction needed to improve efficiency and open new markets. Another threat was the growing competition with Cowles Electric, which was supplying aluminum at 50 cents a pound, far below the market price of 85 cents a pound. Pittsburgh Reduction believed Cowles Electric was selling below cost and was infringing on the Hall patent. At the same time, federal court hearings had started on patent rights between Pittsburgh Reduction and Cowles Electric Smelting and Aluminum Company, and this would continue for years.

Pittsburgh Reduction challenged Cowles on the use of electrolysis and internal heating, while Cowles countered that the idea of internal heating had been stolen by Hall during his period of employment. In 1893, Judge William Taft (future president of the United States) ruled decisively in favor of Pittsburgh Reduction, asking Cowles to pay damages of $292,000. Not surprisingly, Taft had very close ties with the money men of Pittsburgh's Eastside. Thomas K. Laughlin, Director in the Jones Laughlin Steel Company, was the husband of William Taft's wife's sister. Taft was a frequent visitor as president to his supporters in Pittsburgh.

ALCOA lacked major markets outside the steel industry, but the World's Fair of 1893 would change that. With a break from legal problems, the company again looked at developing markets. Many had questioned the Taft decision. In 1933, one reporter noted: "that signature, 'Taft, J.' was worth $100,000,000 to the Mellons."[3] As of today, that dollar amount is well into the billions. Certainly in the future, Taft would find many high-paying donors from Pittsburgh in his presidential campaign. However, in 1893, the dominance of the future ALCOA was still far from assured.

Alfred Hunt and Charles Hall would join Pittsburgh East End exhibitors—George Westinghouse, H. J. Heinz, and Henry Clay Frick—at the 1893 Chicago World's Fair in stealing the show. The Pittsburgh Reduction Company was an award winner at the fair with its educational exhibit. The exhibit included some of the original small pellets made by Hall at his Oberlin laboratory. There was also a scale model of the Hall process and aluminum samples. An aluminum coin that was made for the fair and a souvenir aluminum cigar holder were popular items. The short preparation time limited the number of uses that could be shown at the fair. A year later, at the Pittsburgh Exposition, the company demonstrated an aluminum bicycle and kitchenware such as pots and pans. The potential use of aluminum in bicycles attracted the Wright Brothers, who would use an aluminum crankcase in building their first plane. Hunt started to produce novelty items to promote aluminum, such as aluminum playing cards, comb sets, dinnerware, artistic sculptures, jewelry, inkwells, blotters, letter openers, letter holders, picture frames, and cigarette boxes. In addition, there were more useful products, such as duck hunting boats.

The fair launched an explosion in interest in aluminum products. A Featherweight Blick typewriter made of aluminum, designed by George Blickensderfer Manufacturing, went on sale in 1894. The typewriters were not made entirely of aluminum, but the frame was aluminum. Profits remained low during this period, which probably also contributed to competition. By 1897, the military was using aluminum castings in naval applications. The legal and product development costs, however, continued through the 1890s.

Pittsburgh Reduction was not alone in exhibiting aluminum products at the fair. The Koenig Manufacturing Company of Wisconsin, which at the time was representing a German novelty company, was marketing aluminum combs. The sales success led Koenig into the manufacturing of combs, bicycle guards, mustache cups, and salt and pepper shakers made of aluminum. The success of Koenig would lead to the growth of the aluminum cookware industry in Wisconsin. Captain Hunt wanted to move into aluminum ware stamped out of sheets like steel after seeing the victory of aluminum novelty items.

The Chicago World's Fair of 1893 was the birthplace of the aluminum novelty industry that then led to aluminum cookware. At the time, aluminum was cast, not stamped out of sheet aluminum for such products. Captain Hunt, a practical metallurgist, had tried to develop a market for sheet aluminum as early as 1889. Hunt believed the long-run mass production of aluminum cookware would have major cost advantages. In 1889, Captain Hunt visited Avery Stamping Company, a steel stamping company that produced cooking utensils. Hunt supplied the sheet aluminum to run a trial to produce hotel stew pans. While it proved too difficult for mass production on these machines designed for steel stamping, they were the first aluminum pans stamped in the United States.[4] Hunt found out that steel stamping tools could not be directly converted to aluminum stamping as aluminum flows, much differently from steel in the die setup. It is a similar problem today in the conversion of the automotive industry from steel to aluminum. Hunt started his own research efforts to develop proper machines, since pans made from sheet aluminum could offer a major reduction in price over cast. Sheet cutting for jar caps in the 1890s did develop into a new market as Ball Brothers started using them in canning jars. Soon, product issues were overshadowed by new legal problems.

The Taft decision of 1893 was far from the end of legal problems as two new challenges arose in the 1890s. The first was that of the Heroult process versus Hall. Heroult had not pursued his aluminum process, preferring the production of aluminum bronze. However, on a trip to the United States to expand alloy production, Heroult would meet Grosvenor Lowery, an associate of Thomas Edison. Lowery and Heroult organized the United States Aluminum Company, but it never evolved past the experimental stage. While Lowery had hoped to challenge the Hall patents, Heroult seemed to be content to stay out of the fray. The Heroult patents had won over Hall in England, and the British aluminum industry offered a secure future for him. Interestingly, when Charles Hall received the prestigious Perkin Medal of the Society of Chemical Industry, Paul Heroult was in attendance and even congratulated Hall.

Lowery, however, would team up with the Cowles Brothers for a more serious challenge to the Pittsburgh Reduction and the Hall patents. The Taft decision of 1893 effectively put the Cowles Brothers out of the aluminum business, but they were determined to fight back. In 1892, the Cowles Brothers had purchased some older aluminum patents. These were the patents of Charles Schenck Bradley that were very broad but covered the general Hall patent enough for a legal challenge. It was said that Hall worried about the Bradley patent more than the other challenges.[5]

Captain Hunt during the Spanish-American War, 1899 (Library of Congress).

Charles Bradley was born in 1853 and attended DeGriff Military Institute in New York while studying engineering. In 1883, he went to work for Thomas Edison and performed most of his work at Edison's New Jersey laboratory. Bradley filed and received several patents between 1883 and 1887. The quest for aluminum had been well documented in engineering and scientific journals. Bradley was well versed as to the confluence of aluminum research of the time, and he had patented a process without really proving it out. Hall realized the potential of Bradley's patent to hold up in court. Bradley's patent had correctly predicted the use of electricity in fusing, heating, and electrolysis, which was how the process of Pittsburgh Reduction had evolved. The Cowles Brothers and their Bradley patent would, in fact, be upheld in 1903. The decision was based on the use of internal heating, which had been missing in the original Hall patent. The decision stated: "Patent after patent was introduced claiming new methods of separating aluminum from its ores, but in every instance external fire was used to fuse the bath and maintain it in a fused condition. Many of these inventions were long after the introduction of dynamos…. Indeed, so strongly was the inventive trend toward the employment of external heat that even the defendant's [Pittsburgh Reduction] inventor, Hall, could not be induced to dispense with its use until 1889 … three years after the Bradley application [1886]."[6]

This court loss of Pittsburgh Reduction required them to pay Cowles Electric $3 million in damages. *Engineering News* in 1903 reported it this way: "It will be noted that the above decision does not affect the validity of the Hall patent for the use of cryolite as a solvent for alumina. On the contrary, it more firmly establishes it. Nor can there be any doubt that it was the Hall invention whose successful exploitation made aluminum a commercial metal."[7] The new decision froze production at both companies, forcing a settlement: Pittsburgh Reduction paid Cowles Electric $250,000 for the Bradley patent, and Cowles Electric received a penny a pound royalty on Pittsburgh Reduction's production. If the production exceeded eight million pounds in a year, the royalty increased to a penny and a half. In return, Cowles Electric was to remain out of the aluminum smelting business. The settlement put Pittsburgh Reduction in complete control of the American aluminum smelting and ingot pricing in the United States. Still, European aluminum companies had a cost advantage because of their cheap sources of hydroelectric power.

CHAPTER 6

The Aluminum Industry Comes of Age

BETWEEN 1895 AND 1903, AS THE COURT BATTLES RAGED, Pittsburgh Reduction continued to make major expansions into the production of aluminum. The first major expansion began with the building of a plant to utilize the hydroelectric power of Niagara Falls. The main cost factor of aluminum was in the price of electrical power. Unlike steel, very few workers were required in the smelting process. The building of the world's greatest electrical power station played a large part in Pittsburgh Reduction's market strategy. Niagara Falls was known as white coal. *Harpers Weekly* said in 1889: "There are some things that do not bear description, Niagara Falls and Mark Twain are two of those things." The power of the falls has an almost unbelievable flow of 100,000 cubic feet per second with a peak of 225,000 cubic feet per second. There is a peak period of flow from April to November, and there is even a daily variation of flow, with 9 a.m. being the peak and 9 p.m. the low. German electrical engineer Professor Siemens estimated in 1888: "The amount of water falling over Niagara is equal to 100,000,000 tons [of coal] an hour."[1] Westinghouse Electric would manage the building of one of the world's greatest hydroelectric stations.

Pittsburgh Reduction was blessed with such forward thinking as well. Captain Hunt and Charles Hall kept their focus on long-term strategy, while Andrew Mellon raised capital and supplied the legal resources. Hunt, in particular, saw aluminum as frozen electricity. It was a vision that would guide Pittsburgh Reduction (ALCOA) for over a hundred years. Electricity was truly the heart of the aluminum industry. Even today, the natural resource of hydroelectric power determines the geographical distribution of the industry.

One of the great results of the Niagara power station was the rise of the American aluminum and electrochemical industry. The two biggest customers of Niagara were the Pittsburgh Reduction Company and the Carborundum Company. Within a year, a massive electrochemical industry grew up around Niagara, and sodium, soda ash, sodium peroxide, and calcium carbide plants followed. By 1897, 12,500 of the 13,500 horsepower available for local consumption was for electrochemical plants. Westinghouse enjoyed visiting these plants since he always had a fascination with aluminum metallurgy. The aluminum industry owes a great deal to George Westinghouse and his vision.

The first power came on at Niagara on August 26, 1895. On that August day, "Dynamo No. 2" came online supplying current to the Pittsburgh Reduction Company. Within a few years this cheap power would drive down the price of aluminum to 33 cents a pound. The initial power also went to the electrochemical plants nearby. Dynamo No. 1 didn't come on stream until September 30. Pittsburgh Reduction Company contracted for 5,000 horsepower of the 15,000 available. Because Pittsburgh Reduction needed direct current for aluminum refining, they purchased the turbine mechanical power from the station to run generators owned by Pittsburgh Reduction. In 1896, Pittsburgh Reduction doubled its production to a million pounds per year and doubled that in 1897. This came with the closing of aluminum smelting at New Kensington, which was now too costly. New Kensington would become a product finishing operation. Niagara Falls would change the aluminum industry forever, making hydroelectric power a necessary raw material.

Another big customer of Niagara power was the Carborundum Company. Carborundum was an artificial abrasive of aluminum oxide. These alumina crystals were produced by electrochemical processing like pure aluminum. Larger colored alumina crystals in nature are sapphires and rubies; these fine artificial alumina crystals are extremely hard, making them ideal for grinding and cutting wheels. In many ways, Niagara Falls would represent the foundation of America's industrial dominance.

Pittsburgh Reduction was studying and buying land for extra water power as the Niagara plant was being built. One of these projects was a plant at Massena, New York, using canal power to make and supply aluminum to Canada. The Power Canal, as it was first called, connected the Grasse River and the St. Lawrence River. In that distance, the depth dropped 45 feet and allowed for the harnessing of 200,000 horsepower. Massena is a town in St. Lawrence County, New York, and is located on the northern (Canadian) border of the county. In 1897, construction of the Power Canal was started. The canal was to connect the St. Lawrence River just above the Long Sault Rapids to the Grasse River. When the hydroelectric dam was completed in 1902, the Pittsburgh Reduction Company came to Massena to take advantage of the cheap electricity, and the company began production of aluminum on August 27,

1903. But Massena's remoteness created a labor shortage and Pittsburgh Reduction would have to duplicate its success at New Kensington in community building.

Massena was a small village of former French lumberjacks, named after one of Napoleon's field marshals, André Masséna. Many immigrants came to the Massena area during the Power Canal construction and at the arrival of the Pittsburgh Reduction Company. Originally, there were only 67 employees, but soon Pittsburgh Reduction grew quickly as a supplier for the Canadian aluminum market. The rapidly increasing immigrant work force for the new aluminum plant required housing. In 1904, the company established the Pine Grove Realty Company to manage construction and sale of houses to employees. Workers could buy these houses, which were conveniently located close to the plant, or rent them until rental payments had equaled the purchase price, whereupon they obtained the deed to the property. The company also provided a recreation hall and athletic field; and in 1906, Pine Grove School was built on land given by ALCOA to the Massena School District.

In 1899, Pittsburgh Reduction was also studying Canadian hydroelectric projects as Canada offered many opportunities for hydroelectric plants. Charles Hall was a major supporter of the company developing production in Canada and personally visited potential sites. Canadian hydroelectric power was significantly cheaper because of the sheer volume of flowing water in eastern Canada. In 1901, Pittsburgh Reduction located to Quebec, close to the Shawinigan Falls on the Saint-Maurice River. The Northern Aluminum Company Limited was formed by Pittsburgh Reduction; but in 1925, its name was changed to Aluminum Company of Canada known as "Alcan." Its first plant was in Shawinigan, Quebec, as part of the Pittsburgh Reduction Company, and this Canadian plant would be the root of the world's second largest aluminum company (Alcan). Again, in Canada, Pittsburgh Reduction Company became involved in the development of the town and community. At the same time, Pittsburgh Reduction was exploring sites in Tennessee and North Carolina, with all of the locations being centered on the availability of hydroelectric power.

Pittsburgh Reduction was not alone in using hydroelectric power to produce aluminum; Europe's aluminum industry also depended on hydroelectric. The electrolytic process for aluminum production required large amounts of cheap electricity, which could easily be provided by hydroelectric power in the Scottish Highlands. British Aluminium seized the opportunity and the first aluminum ingots were produced at Foyers, Scotland, in the highlands in 1895; and its first hydroelectric-powered smelter opened in 1896. British Aluminium was the only other fully integrated aluminum company at the time. The French and German industries were sectioned into primary producers and fabricators. Scotland became part of a craze toward aluminum as the price came down, and aluminum became a popular metal in many new applications. In 1895, the Scottish shipyard, Yarrow & Co., built a 190-foot-long aluminum torpedo boat for the Russian navy, *Sokol*, that attained a speed record in its day of 32 knots. Also in 1895, the aluminum yacht, *Defender*, won the America's Cup. At this time, British Aluminium and Pittsburgh Reduction saw the opportunities as unlimited.

The combination of vertical integration, technology, and tariffs protected Pittsburgh Reduction from serious competition in the United States. Pittsburgh Reduction had also aggressively purchased and created exclusive rights to all American sourced bauxite operations, breaking all dependence on Europe. The real key to Pittsburgh Reduction's growth was with its strong team. Charles Hall became the company scientist working on process improvement, and Arthur Davis became the operations manager. Captain Hunt proved to be an outstanding marketer, and his passing in 1899 hurt the company's growth. The Mellons supplied the financial wisdom and, of course, they supplied the first-rate legal support as well. The Mellons' interests also helped in a strategy to pay little dividends, while pouring earnings into expansion, following in the footsteps of fellow Pittsburgher Andrew Carnegie. These men kept the focus on the future with long-term strategies through the daily legal and production problems.

Dynamos and pot room at Smallman Aluminum Works (Library of Congress).

One of those long-term strategies was vertical integration. Pittsburgh, with the Carnegie, Westinghouse, and Heinz empires, was the center of vertical integration. Vertical integration means control and ownership of the raw material supply chain (backward) and also end product production (forward). H.J. Heinz had actually studied vertical integration before Andrew Carnegie, who is often credited with its successful application.[2] For Pittsburgh Reduction, vertical integration would become necessary for success and growth. In its earliest years, Pittsburgh Reduction purchased a pure refined alumina and did not use bauxite directly. Pennsylvania Salt Company acted as an agent and supplier. At the time, one half the cost of producing aluminum was that of the purchased alumina.

With the huge amounts of energy available at Niagara Falls, Charles Hall and Captain Hunt looked to other means to reduce costs. In 1889, Hunt wrote the following to Charles Hall: "If it be true, as I understand from you, that we can make as much aluminum per unit of horsepower from bauxite as from pure alumina, then it would seem that this is the field which is now ripest for us to investigate for increasing the economy of manufacture."[3] Even back at Oberlin, Hall had been experimenting with ores such as bauxite, believing even in the potential of common clay as an ore. Control of cheap bauxite ore and hydroelectric power would make Pittsburgh Reduction the leader in the international aluminum industry.

Furthermore, in 1897, Captain Hunt set the strategy for the country: "Looking forward to the time the Hall patent will have expired, and it must be our policy now for the next few years to strengthen and solidify the position of Pittsburgh Reduction Company that we shall be independent of both the tariff and patent situation. The way to achieve such independence was through continuing expansion of the company's fabricating facilities, on the one hand, and taking greater control of electrical power and mineral resources on the other."[4] In hindsight, this has been the guiding policy of ALCOA throughout its history.

In the early 1890s, bauxite was found in Georgia and Arkansas. John C. Branner, Arkansas state geologist, was the first to identify bauxite when he noted it in a sample brought to him by Ed Wiegel of Little Rock in 1887. The material was being used to surface the roads from Sweet Home to Little Rock. Branner's first published report of the occurrence of bauxite appeared in 1891. In Arkansas and Georgia, an early deposit was being mined by Georgia Bauxite and Mining Company. The New Kensington plant had been operating as an experimental plant for bauxite under Charles Hall's supervision since the opening of the Niagara Falls operation. It was here that a small amount of Georgia/Arkansas bauxite was successfully processed in 1899. Pittsburgh Reduction continued to increase its use of Georgia/Arkansas ore, and in 1900, it purchased Georgia Bauxite and Mining Company. In 1903, Pittsburgh Reduction built its

St. Louis plant to process the bauxite from Georgia and Arkansas. St. Louis was selected because of its position near cheap Illinois coal, which was utilized to process the bauxite. Pittsburgh Reduction was able to greatly improve upon their process for bauxite at St. Louis, but it was far short the efficiency of the competing European Bayer process. Still, Hall was determined to meet the challenge.

In 1905, Pittsburgh Reduction also purchased the General Bauxite Company of Arkansas. The Arkansas deposit was 22 miles south of Little Rock and had shipped bauxite to Pittsburgh Reduction since 1899. Pittsburgh Reduction was again faced with the need for community building, so they built their own community called Bauxite to support its mining operations. Pittsburgh Reduction started with nothing and built a town that was a true company town, modeled after Pennsylvania mining towns with improved views on the needs of the workers.[5] The Mellons were particularly sensitive to claims of company town abuse from their earlier investments in Pennsylvania coal mining towns, and they wanted it done right from the beginning. Of course, their motives weren't completely altruistic; the mood of the country was very anti-business, with a populist belief that big companies took advantage of their workers. Coal mining and steel towns were often modern subsistence living. Pittsburgh Reduction followed the Westinghouse model of building community versus mere housing. Housing at Bauxite was supplied on a rented basis, as well as community centers for religious and recreational activities, and the company would provide medical facilities and schools as the community grew. In general, Pittsburgh Reduction, and later ALCOA, maintained a high standard of living for this mining town and had very favorable press. Bauxite grew as Pittsburgh Reduction grew.

Even with the St. Louis plant and the southern mines, Pittsburgh Reduction was short on refined alumina and continued to purchase alumina from Pennsylvania Salt Company. While the use of refined bauxite did cut costs, the Pittsburgh Reduction operation was inefficient and expensive compared to the Europeans. The problem was, the best refining process was that of German Karl Bayer, whose patent was good until 1911. At that time, the company (ALCOA) switched to the Bayer process, which started with bauxite of 55 percent or more aluminum oxide and less than 7 percent silica. The bauxite is crushed and heated in a caustic solution, and this produces aluminum hydrate in solution and a waste product called red mud. The aluminum hydrate is then precipitated and washed. The process requires large quantities of water, soda ash, and lime. ALCOA had been improving the process for years, but alumina production would remain a bottleneck for decades. Pennsylvania Salt was also Pittsburgh Reduction's supplier of cryolite, and they needed to make a smooth transition.

Cryolite was every bit as critical as bauxite itself. With all of Pittsburgh Reduction, cryolite remained a small but vital link in the chain. Cryolite came from a single supply in a Danish colony on Greenland and was also an opportunity for vertical integration. At the time, all cryolite was in the control of European mining companies, and Pennsylvania Salt Company had exclusive American rights to cryolite. The supply and availability of cryolite was limited by the mine location in the Arctic Circle. The frozen shipping routes and limited ore ships allowed for only two rounds per year. Germany and France had been producing synthetic cryolite by 1900, but Pittsburgh Reduction had to struggle for another decade. The problem would reach the crisis stage for Pittsburgh Reduction as war swept Europe in 1914, and supply was threatened.

CHAPTER 7

Metallurgical Wars and Monopoly

AT THE GOLDEN JUBILEE OF ALUMINUM 1936, Arthur Davis, then CEO of ALCOA, noted the four epochs of the company: "(1) Can we make aluminum? And this we were able to answer in the affirmative as soon as our production reached 30 pounds per day. (2) What can we do with what we made? Which became an early problem as our output, small as it was, piled up on our hands and was answered by making novelties of it. (3) Can we make any money on it? This was finally answered by our going into the business of doing our own fabricating. (4) How can we make the business grow? This still keeps us searching actively for new materials through research, despite the fact that our present production is in the neighborhood of 300,000,000 pounds per year [1936]."[1] The epochs were a history of Pittsburgh Reduction's offensive strategy. The common factor for all these epochs was the continuing legal struggles.

The outstanding management of Pittsburgh Reduction had steered it through difficult legal challenges as well as raw material supply issues. The management had looked forward in the development of new markets for aluminum and poured earnings back into company growth. Growth was steady, but investment costs were high as well. However, Pittsburgh Reduction became the dominant single producer of aluminum, and only British Aluminium could challenge them in the world market. This dominance had come in an environment of trust-busting in other industries in the United States. For the first decade of the 20th century, Pittsburgh Reduction was viewed as a legal monopoly—legal in the sense that it was operating under patent protection until 1909. After 1909, the company would fall under the scrutiny that had been focused on oil and steel industries.

The Aluminum Company of America (ALCOA) became the firm's new name on January 1, 1907. By that year, the company operated hydropower and reduction plants in Niagara Falls, New York (1895), and Shawinigan Falls, Quebec (1900); mining operations in Bauxite, Arkansas (1901); and reduction facilities in East St. Louis, Illinois (1902). ALCOA's growth was aggressive but within the legal restraints of the period. ALCOA had made arrangements to prevent its own supply of primary aluminum from going into competition with it. Fabricators and rolling mills were all dependent on buying primary aluminum ingot from ALCOA. In 1912, the federal government would challenge many of the arrangements Aluminum Company of America (ALCOA) had made with other companies to keep them out of aluminum production.

For a decade, ALCOA grew in the shadow of large trusts in industries like steel, oil, tobacco, cotton, and pig iron. President Teddy Roosevelt in 1903 instructed his attorney general, Philander C. Knox, to launch a series of lawsuits against what were deemed offensive business combinations. Such giants as J.P. Morgan's Northern Securities Company, John D. Rockefeller's Standard Oil Trust, and James B. Duke's tobacco trust, were targets of the government's attorneys. Philander Knox was a Pittsburgh eastsider and a poker buddy of the founders of ALCOA and United States Steel, which may explain why they escaped in the initial campaign on trusts. The basis for the action was the Sherman Act of 1890.

The Sherman Antitrust Act prohibited certain business activities that federal government regulators deem to be anti-competitive and requires the federal government to investigate and pursue these questionable trusts. Any combination "in the form of trust or otherwise that was in restraint of trade or commerce among the several states, or with foreign nations" was declared to be illegal. Trusts or monopolies cut prices drastically in order to drive competitors out of business. Among their other anti-competitive techniques, trusts were buying out competitors, forcing customers to sign long-term contracts, forcing customers to buy unwanted products in order to receive other goods, restricting interstate trade, entering into unfair pricing arrangements, and joining foreign cartels. Under patent protection, ALCOA's control of primary aluminum price and who was supplied was deemed valid. ALCOA's cartel arrangement with British Aluminium, however, garnered more scrutiny even with the patent protection.

ALCOA was clearly participating in illegal cartels in 1908 and the years up until World War I. Andrew Mellon became personally involved in ALCOA in the early 1900s. He promoted aluminum products in his home and businesses. Aluminum became his passion, and he used his international banking connections to make ALCOA an international company. He personally negotiated a business arrangement with British Aluminum in 1907. The cartel between the two allowed for British Aluminium to take three million ingot tons per year from ALCOA. In return, ALCOA would not sell finished aluminum in England. Furthermore, the two would not sell to the other nation's government. This was clearly an illegal arrangement in violation of the Sherman Antitrust Act.[2] A year later, Mellon set up an arrangement with the German amalgamation of European aluminum companies known as Aluminum A.G. The arrangement divided up the European market and other markets in Africa and Asia.

By 1912, ALCOA became a potential target of the government. ALCOA's issues involved its restrictive supply arrangements with domestic companies like Pennsylvania Salt and Norton Company. Another issue was its membership in a European cartel. The European companies competed fiercely with one another at first, and then signed a series of cartel agreements after 1896. In 1901, the first worldwide aluminum cartel was formed by the five major companies who were producing 90 percent of the world's aluminum at the time. The cartel had the five significant early aluminum companies at the turn of the century: one American (ALCOA), two French, one Swiss, and one British. Each company would be free to sell within its own country in a closed market, at a price around one cent a pound higher than the world price. Then each company was allotted a guaranteed share of the rest of the world market (ALCOA, as Northern Aluminum of Canada, received 21 percent). As a result, all the firms were able to charge around 36 cents/pound in Europe (instead of the 22 cents they had been receiving); meanwhile, ALCOA charged 33 cents/pound in North America for a few years to stimulate demand, then charged to the cartel a price of 36 cents/pound from 1906 to 1908, by which time ALCOA was making an annual net profit of 35 percent on stockholder equity. In general, the aluminum industry operated free of the hard public scrutiny given to other heavy industries. ALCOA had, for the most part, been free of national labor problems, and this contributed to its good public image. Part of the reason ALCOA had been free of labor issues is that aluminum smelting is not very labor intensive compared to steelmaking or auto production. In addition, labor costs were a small cost factor after electricity and bauxite.

The government review of ALCOA came under the favorable administration of President William Taft. The legal challenges and investigations of ALCOA and other companies were also slowed by the onset of World War I. The proceedings ended in 1912 with the government issuing the Consent Decree of 1912, and ALCOA agreed to stop all questionable practices. Actually, the main changes came a few months before the Decree of 1912. The contract with Pennsylvania Salt was terminated, the cartel arrangement ended as Europe moved into a war environment, and afterward, ALCOA emerged all the stronger from the legal review. Market and technical challenges would be the greater test for the balance of the decade.

The Decree of 1912 opened up the possibility of foreign competition building production plants in America. French producers had started a plant in North Carolina in 1915, before the war intervened and shut down the project. Tariffs were reduced two cents a pound after the Decree of 1912. Imported ingot aluminum did increase to as much as 30 percent (from less than 2 percent) before the war. Once the war started in Europe, imports quickly dried up.[3]

During the struggles of 1912, ALCOA continued to expand new consumer markets. Prior to 1912, the major use of aluminum was as a deoxidation agent for steel. Cooking utensils had been an area of growth in 1912 and, by 1917, accounted for approximately a third of ALCOA's production. Another growing market was aluminum cable, which ALCOA had made since 1898. European aluminum was aggressive in entering the American markets. Congress, however, was determined to make aluminum America's industry and placed tariffs on finished aluminum product. ALCOA, for its part, poured profits into building plants and increasing employment. Like the original McKinley Republican scientific tariffs, Congress oversaw protected companies to assure profits were put into expanding American employment. ALCOA also aided in the development of an aluminum consumer goods industry in new areas. Besides its own Wear-Ever division, in 1909, ALCOA had invested in the cookware company called American Goods Manufacturing. The revolutionary cookware changed the American kitchen forever because of its resistance to rusting, its remarkable weight advantage, and its seeming ability to "wear-for-ever." Aluminum Wear-Ever cookware became world-renowned in 1909, when Admiral Robert Perry took the cookware on an expedition to the North Pole. By 1912, ALCOA controlled 75 percent of the aluminum cookware market.

Still, the cookware market was a real challenge for ALCOA and others because the housewives needed to be educated and trained on the use of aluminum cookware. Aluminum heated up three times as fast as iron and steel, but the heat was not as even as in the old cast iron skillet. This difference was less noticeable in heavy cast aluminum than the sheet metal aluminum, but it required new cooking skills. ALCOA had to invest in a major advertising campaign and become active at local exhibits. ALCOA pioneered door-to-door sales of aluminum cookware to help directly demonstrate cooking with aluminum. While ALCOA was in control of the market, there were aggressive challenges by foreign producers.

Part of the success of imported aluminum cookware before and after World War I was due to the thinner sheet metal used by German manufacturers. Thinner gauge greatly reduced the cost of the cookware. ALCOA had maintained thicker gauge aluminum sheet was needed to maintain the quality of the finished product. Still, the Germans successfully targeted a low-price segment of the market. As early as 1901, ALCOA had promoted the formation of the American Aluminum Association to assure quality in the marketplace, since aluminum cookware was in competition with a number of other materials such as steel, cast iron, and copper. In a 1920 report to the Federal Trade Commission, the following was noted: "The average housewife in the country knows nothing about the difference in gauges [thickness] or the value of aluminum cooking utensils. All aluminum looks alike to her. She does not know that a heavy gauge utensil is invariably better made, more serviceable, less liable to injury, and will stand wear and tear to constant use in the kitchen better than a light gauge one. The retailer usually makes more money on the light gauge and tells his customers it is just as serviceable."[4] The market problem forced the use of association standards and educational advertising. Eventually, ALCOA and other American cookware companies created two quality brands. One brand made lighter gauge to meet the price pressure of cheap imports, and the other brand was top of the line with heavy gauge. ALCOA was defensive on many market fronts.

With the great Niagara Falls power plant, there was a need for cable to transmit AC electricity long distances. Transmission cables were originally copper, but the copper was heavy and expensive. Pittsburgh Reduction had starting promoting aluminum cable as a substitute, and electrical engineers were hired to help with the design. Metallurgists used a two percent copper additive to add strength, and the actual wire was reinforced with steel wire. Pittsburgh Reduction built a wire mill at New Kensington to make aluminum wire. While aluminum was more expensive and less conductive, it substantially reduced the need to construct support towers. By 1900, aluminum cable was the preferred product in transmission lines, and ALCOA owned the market.

ALCOA's production was outpacing demand, as was true with the European aluminum companies as well. Research on aluminum in America was driven by private industry and consumers. New markets needed to be developed and expanded with the help of ALCOA's research and expertise. Aluminum bicycles had been around since 1892, and the Wright Brothers had been working with them, too. The Lu-Mi-Num bicycle frame was cast hollow in one piece as, at the time, there were no sheet mills for aluminum. ALCOA saw the need and started its own sheet mill at the New Kensington plant. Later in the 1890s, Cycles Aluminum of France bought the rights to the Lu-Mi-Num. The interest in aluminum was probably at its highest as cast aluminum was being used in motorcycles and sidecars. By the turn of the century, aluminum's future would be in transportation. Bicycles would take the aluminum industry to the newest form of transportation along with the automobile and the emerging aircraft applications. The Wright Brothers, who were in search a lighter engine, turned to aluminum in 1903.

In the 1890s, with the aluminum bicycle, aluminum moved to horse-drawn carriages. The *New York Times* reported recently: "Studebaker, once the world's largest maker of horse-drawn carriages and buggies, according to the Studebaker National Museum in South Bend, Ind., built a farm wagon in 1893 whose metal components were largely aluminum."[5] The wagon contained 149 pounds of aluminum, and it took more than 4,000 hours of labor to build and cost $2,110.65, a hefty premium over the $200 cost for a typical farm wagon. It was a short step from wagons to automobiles.

Earlier work at ALCOA's New Kensington plant consisted of aluminum castings for auto engines to reduce weight. In 1904, New Kensington found success in the casting of auto bodies, prior to the Ford Model T, when there were over 200 independent car companies. It would be in 1914 that the Model T's use of steel, ironically, with small aluminum additions to remove impurities, allowed for the stamping of steel sheet. The welding and formability and low cost made Ford's mass production work. Engine parts, however, remained a major and expanding market for aluminum and would take the industry into the new market of aircraft.

The Wright engine was cast in Dayton using aluminum from Pittsburgh Reduction. They were aware of the use of aluminum engine parts in German race cars as well as American cars. The four-inch bore, four-inch stroke, cast-iron cylinders fit into a cast aluminum crankcase that extended outward to form a water jacket around the cylinder barrels. Buckeye Irons and Brass Works of Dayton and the Wright Brothers selected an alloy of 92 percent aluminum and 8 percent copper. The engine was based on German experiments and early American work with copper as an additive. The Wright engine, with its aluminum crankcase, marked the first time this breakthrough material was used in aircraft construction. The Wright Brothers' first flight brought aluminum back into the spotlight. The military was a bit slow in pushing for potential applications, but aluminum would eventually be used in early propellers for the military.

The success of the Wright brothers' engine opened a major market in automobile engine parts as well. By 1914, 80 percent of auto crankcases and gear cases were aluminum.[6] While the number of aluminum car bodies declined, the market for other auto parts boomed, including rear axle housings and engine beds. Additionally, some aluminum sheet was being used in cars, too.

ALCOA's sheet mill opened the auto body market again because sheet led to a major reduction in aluminum needed. By 1915, 65 percent of all new aluminum went into automotive parts. Aluminum weight reduction offered speed and fuel savings, albeit at a cost over wood and steel. The first sports car featured a body made of aluminum and was presented to the general public at the Berlin International Motor Show in 1899. Mercedes Benz also constructed the first aluminum car engine in 1899. King Leopold of Belgium ordered an aluminum car from Mercedes in 1902. Pierce-Arrow Motor Car Company started to cast auto bodies in 1904. The light weight and corrosion resistance made it popular in such a luxury car. Rolls-Royce's Silver Ghost, produced in 1907, was a luxury car with an aluminum body. Some small American and German auto producers were using aluminum sheet in car bodies as early as 1910. Alfa Romeo's Aerodynamica in 1914 was a beautifully designed type of minibus that would augur the aluminum Dymaxion 8-seater of Bucky Fuller in the late 1920s. The cost and technology of aluminum sheet prevented the market from expanding to the mass production of Ford Motor. The auto industry, prior to World War I, drove the American and ALCOA's research effort, but Ford's mass production presented new challenges. To meet the challenge, ALCOA invested heavily in research and development.

In the early 1900s, research in aluminum was much different in Europe than America. ALCOA was focused on process cost reduction and consumer markets, while Europe looked to its military uses. The European research, in particular, focused on the emerging use of aircraft in war. Furthermore, European countries preferred a government-industry alliance with the military taking the lead. Ferdinand, Graf [Count] von Zeppelin (1838–1917), was a German general and later an aircraft manufacturer who would change the face of aluminum research. In the early 1890s, Zeppelin wanted to use aluminum in airships before production amounts of aluminum existed. German Zeppelin airships were built in the early 1900s using wood as the framing to hold hydrogen gas bags. Zeppelin was experimenting with aluminum, but it lacked the strength and rigidity needed, even when it became commercially available. Zeppelin aircraft became the center of German military research focus, and aluminum offered a major weight reduction over wood. The German military immediately saw potential in a lightweight airship for bombing, but a new aluminum alloy would be required. The militaristic German government was willing to invest heavily into aluminum research, and a new leader emerged to champion aluminum.

In the 1870s, Napoleon III had championed the potential use of aluminum applications in war. In 1902, a new patron of aluminum arose in Europe named Wilhelm II, King of Prussia and Emperor of Germany. Wilhelm II commissioned his scientific research center at Neubabelsberg, southwest of Berlin, to research and develop alloys of aluminum for military applications. German metallurgical engineer Alfred Wilm (1869–1937) was assigned to the project of developing a strong lightweight alloy of aluminum. Europe had advanced much further in the engineering field of metallurgy. In the Prussian wars of the 1870s, metallurgy had proven to be the key to victory using advanced armaments. Germany had organized its research into an industrial-military complex capable of fast-tracking new weapons. England and France also had similar arrangements between the military and private industry.

While Hall and Heroult are always linked with the commercialization of aluminum, without Wilm, aluminum would have been limited to jewelry and steel production. When we think of aluminum today, it's really closer to Wilm's Duralumin than the pure aluminum of Hall and Heroult. Aluminum had the lightness factor, but lacked the strength, unless designed in bulky castings. It would be Duralumin and its derivatives that changed aluminum into the titan of metals. This alloy of Duralumin became the backbone of the aircraft industry and it would open the door for aluminum in architecture, building, and automotive parts. It would make aluminum, like steel, heat treatable to higher strength and hardness. Even more amazing is that Wilm's Duralumin would be the base of today's super light aluminum–Lithium alloys.

French artisans, prior to the Hall-Heroult process, had learned that an alloy containing two percent copper produced a superior metal. Copper, when added to aluminum, improved hardness, was easier to engrave and didn't dent easily. German metallurgists had developed an aluminum-copper alloy in 1902, which the Wright brothers used in their engine. By 1906, Alfred Wilm had developed an alloy of aluminum using copper, manganese, and magnesium additives. The actual composition was copper (1.3 percent), magnesium (2.8 percent) and manganese (1 percent). Wilm used heat treating to develop the strength and rigidity in the alloy. It was known as alloy 2017, but in today's terminology, it is the basis of the 6000 and 5000 series. This new experimental alloy was named Hartaluminium originally. Wilm was reassigned to German metal factories in the city of Duren to perfect its commercial manufacturing in 1908. In was in Duren that the alloy picked up a new name—Duralumin. Wilm had applied for a patent in Germany and England and the patent was particularly vague to limit duplication of the metal and process. England and Germany, at the time, were both ruled by dynasties of German blood, allowing for some technology exchange. The British soon became fearful of the use of Duralumin to build Zeppelin airships capable of massive long-distance bombing runs. During World War I, the Germans would produce Duralumin for 80 airships—more than 726 tons in one year.

The discovery of Duralumin rivals the discovery of steel. Duralumin and steel are the two alloys that could be said to have changed the world. Duralumin clearly revolutionized metallurgy and technology. It might even be said that we moved from the Steel Age to the Duralumin Age in 1919. The metallurgy of aluminum heat treatment behind Duralumin is the foundation of the widespread use of aluminum in space, high-tech products, aircraft, sports equipment, and building today. Wilm had given a laboratory technician the assignment of testing the physical properties of various Al-Cu-Mg [aluminum-copper-magnesium] alloys. Since the weekend was approaching, the technician left the samples unattended until the following Monday. On Monday, he found unusually high strength in the samples, which was a result of the Duralumin sample's having been stored for two days at room temperature. This process is known as "natural aging."[7]

Aging is a process by which internal stresses grow with time and temperature to harden and strengthen alloys. Wilm soon learned to use increased temperature treatment to accelerate the aging or strengthening. However, one had to be careful as to not over-age Duralumin, which would make it brittle and subject to failure in applications. The heat treatment and aging process make up the real secret of the discovery. It allowed a part to be formed while the Duralumin was "soft," then using heat treatment to harden and strengthen the part. This was a breakthrough in metallurgy. The Germans learned to make aluminum rivets and then heat them to connect aluminum sheets with the hot rivets. The rivets not only hardened, but "grew" slightly, making a very tight joint. The Duralumin technology was one of Germany's most guarded secrets.

The British were able to purchase Duralumin from Germany as well as produce it under license. The British armaments company, Vickers, was commissioned by the British military to work on an aluminum airship in 1908. The British moved faster than Zeppelin in producing an aluminum superstructure for an airship. The actual engineering design of the German Zeppelins was obtained through industrial espionage.[8] Vickers Company favored wood for the structure, but the British Admiralty insisted on aluminum. By 1911, the British beat the Germans to flying an aluminum dirigible with the *Mayfly*. However, the *Mayfly* was plagued with problems because the use of unalloyed aluminum lacked the strength, and their copy of Duralumin was imperfect. The *Mayfly* was damaged and had to be scrapped on one of its first long trips, a major setback for Vickers Company. In the meantime, the Germans were making steady progress on an aluminum dirigible.

The Germans had successfully been flying wood-frame dirigibles since 1900, but they slowly substituted aluminum in the frame with the development of Duralumin. Their first all-aluminum (Duralumin) frame dirigible, *LZ-26*, successfully flew in 1914. The Germans had perfected the rolling of Duralumin sheet to be made into girders, and soon their success created fear throughout

Europe. England had turned to steel framing, which proved unsuccessful as steel was just too heavy. When war broke out in Europe in July of 1914, England returned to aluminum research and development. However, the Germans broke off their cooperation on aluminum, leaving the heat treatment of duralumin and its processing a secret.

The war in Europe also motivated the U.S. Navy to turn to ALCOA for the building of an aluminum dirigible, but ALCOA was too far behind on alloy development and fabricating technology for such a project. Most of their alloy work had been under Captain Hunt in an effort to produce a wear-resistant horseshoe. The ALCOA metallurgists really had no idea as to the nature of Duralumin. ALCOA had some alloys of aluminum using copper and zinc, but they lacked the strength of Duralumin because they did not have the technology of aging. In 1916, they obtained some samples from a downed Zeppelin in France. ALCOA then worked with the British and their espionage to find the formula for Duralumin, but it took captured German technology after the war to unlock the secrets of the aging heat treatment. The problem was that they needed to have the composition and the complex heat treatment to gain the full benefits. ALCOA finally produced a similar alloy known as 17S (known today as 2017-T4) after World War I, which could be heat-treated to the same strength as Duralumin. They set up an experimental rolling program at New Kensington to manufacture Duralumin sheet by 1919, and were finally successful in producing an American dirigible in 1923 called the *Shenandoah*.

Germany proved to be the largest user of aluminum during World War I. Thanks to their alloy development and heat treatment technology, Germany used aluminum more extensively in the manufacturing of cars, trucks, and tanks. In 1904, Germany consumed 4 million pounds; but in 1913, at the eve of the war, it was consuming 20 million pounds. By the end of the war, Germany was consuming over 75 million pounds. The German's Duralumin met all the requirements of master aircraft designer Hugo Junkers, which included high strength, forgeability, and the incredible lightness for a metal. Originally, the Junker J2 was constructed with steel sheeting; but it was too heavy, and the dense steel weighed the aircraft down, so the company then experimented with Duralumin, which became the standard for metal wings. While the early tests were initially successful, new metallurgical problems arose. The welding of Duralumin, like steel, failed as the seams quickly corroded.

The Germans then developed specially engineered rivets and company-designed pneumatic tools that revolutionized aircraft manufacturing. In 1917, the Junkers 7 fighter, entirely built of Duralumin, took off from Adlershof airfield. In the same year, production of Junkers military airplanes was started; they were ordered by the German Ministry of Defense for participation in World

War I campaigns. The Junkers airplanes were named "flying tanks," and the Duralumin reliably protected the pilot from bullets and shells. There is a recorded case in which Duralumin sustained 480 bullet shots on the wings and fuselage, and the airplane not only completed the combat mission, but also successfully landed at base.

German war research of aluminum had paid huge dividends. Just as important as Duralumin itself was the Germans' development of extruded tubes and beams to support their Zeppelins. The 1930s German Zeppelin program, like the space program of the 1960s, expanded technology in all phases of science and engineering. The Germans developed rolling sheet technology needed in the Zeppelin program, while America focused on cast aluminum parts. The Germans had been producing thin-gauge aluminum sheet since 1904. Thin sheet had opened new markets for cookware and furniture, and Zeppelins, in general, promoted an array of furniture and light tables. The Germans learned how to do everything using high-strength Duralumin. American extrusions, forgings, and tubing had been developed around soft alloys by engineers not knowing the technology of Duralumin. These softer extrusions limited their use in frames for airships and bicycles. Lacking the Duralumin technology, ALCOA and American aluminum manufacturers expanded their efforts into aluminum casting technology.

The American effort during the war did advance the use of aluminum in cast aircraft engine parts. The Liberty 12-cylinder engine made substantial use of aluminum, including aluminum pistons; however, the German aircraft engines, such as Mercedes and Duesenberg, had more aluminum in their engines, and the average engine had about 300 pounds of aluminum. ALCOA was also able to introduce aluminum skin in the wings of the famous Curtiss Jenny in 1918, but it lacked the strength and bullet resistance of German planes. ALCOA was also able to produce gas tanks, seat backs, and other frame parts for aircraft, although without an alloy like Duralumin, American industry was at a disadvantage. Still, the demand for Allied countries' aircraft aluminum was over 200 million pounds for the war. ALCOA production went from 109 million pounds in 1915, to 148 million pounds in 1918. ALCOA moved all its operations into supplying the government as it struggled to meet orders. The company was also exporting aluminum to our allies.

One area in which America had an edge in research and development was with sand castings of aluminum. Sand casting offered a way to produce car bodies, bicycle frames, and various engine parts; and while castings were thicker and heavier, they could meet strength requirements. In 1909, ALCOA became a member of a five-foundry corporation, the Aluminum Castings Company, to research and develop sand castings. The company proved short-lived when, in

1916, ALCOA bought back its New Kensington foundry as a military supplier. However, the expertise would pay off in the 1920s. The oncoming war increased both aluminum castings and general aluminum parts. Foundries and companies around the world found themselves short of strategic primary aluminum.

Initially, Europe turned to America to supply the enormous demand. The major demand for aluminum came from its use as a powder mixed with ammonium nitrate for explosives. Aluminum powder had been developed by Charles Hall in 1908. Aluminum powder added to paint gave the paint anticorrosive and protective properties. A plant to produce powder was built in Dover, New Jersey, in 1910. By 1914, most of the powder was going to military demand. ALCOA converted its New Kensington plant to all-military applications, basically eliminating its production of cookware. Also, ALCOA built a new smelting plant and nitrate plant in Alcoa, Tennessee, to help meet this European demand. From 1915 to 1917, British Aluminium had purchased as much as 70 million pounds of aluminum powder. The demand created a major shortage of bauxite from Arkansas; and by the end of the war, ALCOA was building a huge bauxite plant in Baltimore, Maryland, to process South American ore. To transport the ore, ALCOA contracted for the building of a number of large steamships, and they added new shipping terminals on the Mississippi River. In an effort to increase hydroelectric power, ALCOA started a power plant on the St. Lawrence River in Massena, New York. The period from 1910 to 1920 saw a boom in the demand for aluminum that often went unmet.

As the decade of the 1910s progressed, Europe found itself short of bauxite. In 1912, a group of French aluminum makers, headed by Pechiney, started to build a U.S. aluminum factory in North Carolina to use the newly found American bauxite deposits. The company was known as South Aluminum Company. The company hoped to combine cheap American energy with rich French bauxite ore. The plant was one of the largest in the world at the time, and the investors hoped to develop a town named Badin. The town was named after Adrien Badin, who succeeded A.R. Pechiney at the head of the company as managing director, from 1914 to 1917. Power for the operation was supplied by a hydroelectric station the new owners built at a location on the nearby Yadkin River known as the Narrows. At the Narrows, the Yadkin River flowed through a deep gorge for three and a half miles within the Uwharrie Mountains. The town was started in 1913, and today still show signs of French design.

In 1915, the Aluminum Company of America purchased the unfinished aluminum smelting complex of Pechiney in central North Carolina from a failing French firm. Two years later, the company had completed the plant, a dam, and a hydroelectric plant, had developed the town of Badin, and began to produce aluminum. The Badin plant would be a major supplier of aluminum for

Left to right: **Richard King Mellon, Andrew Mellon and Paul Mellon, 1929 (Library of Congress).**

the Allies during World War I, and this North Carolina plant gave ALCOA a surplus of primary aluminum.

ALCOA also expanded its fabrication plants and teamed up with foundries during this prewar period. ALCOA opened its Cleveland plant in 1909, and it would supply aluminum castings for the military applications in engines. ALCOA greatly expanded its finishing operations during the war. Canteens, field mess kits, and dog tags were other American military applications in World War I. ALCOA's expansion was massive in terms of ingot aluminum in order to meet war demands, but it lagged behind in the production and development of high-strength aluminum alloy and rolling mills. Demand for tonnage had put limited research resources into the production of primary ingot and bauxite processing. They had also lost their research champion in Charles Hall in 1914.

On the eve of war in the year 1914, the aluminum industry would lose two titans in Hall and Heroult. At the time of Hall's death, the industry was continuing to grow, but was in search of new products to create from the surplus of ingot. From the time of Hall's birth to his death in 1914, the price of aluminum went from thousands to 18 cents a pound. Hall's passing would be honored around the world. Because Hall never married and had little family, over the years he had become very close to his partner, A.V. Davis. Hall's will gave A.V. Davis his stock and the voting rights to additional stock at Oberlin College. Hall had become a member of the board of trustees of Oberlin College in 1905, and Oberlin was left stock in the company, making it a part owner. Davis became president, then chairman of ALCOA, and moved to New York City. Roy Arthur Hunt, Captain Hunt's son, was made operating manager and day-to-day administrative manager in Pittsburgh. Davis's brother, Edward, was in Pittsburgh, and worked on developing ALCOA's sales force. All of these major changes came as a result of Hall's death.

Hall died in Daytona Beach, Florida, on December 27, 1914, at the age of 51. The price of aluminum was 18 cents a pound on that day. His competitor and friend, Paul Heroult, had died on his yacht six months earlier. Hall left Oberlin College approximately five million dollars in his will,[9] and another significant part of his bequest went to Berea College in Kentucky as well as other educational enterprises at home and abroad. The Berea College was considered the "Oberlin of the South" and was known, like Oberlin, for its interracial approach to education. Hall also left money for education institutes in China and Asia in honor of Oberlin missionaries killed there during the Boxer Rebellion. For most of his life, Charles Hall lived in his laboratories, never married, and lived very frugally with extremely limited interests outside of chemistry. Unlike his partner Captain Hunt, he published very little and was not active in engineering societies. He satisfied his interest in concert music and opera by attending performances in the great halls of the United States and Europe as he traveled for the company; and he always had a rented piano in his homes. He attended church regularly and drew strength from the stories of great men who sacrificed for their convictions. And he developed a collector's eye for fine Oriental rugs and porcelains.

Hall's death also marked a change in the methodology of research. The war had created higher demand for aluminum and the creation of a unique set of products. ALCOA started to hire metallurgists and electrical and design engineers to limit the trial-and-error experimental approach of the founders. Proven scientific theory could be used to focus research, cutting the time and number of experiments. Even so, the early research centered on Charles Hall versus corporate structure. His methodology was experimentally based with less theory, unless verified by experiment. In addition, Hall's research focus was on the reduction of cost in primary aluminum production versus alloy development.

The European industrial government alliance was lacking in the United States, and ALCOA was far behind Europe in alloy research and development. President Woodrow Wilson had envisioned government-focused research as in Europe, which would develop military applications. Wilson, at the onset of war, nationalized the railroad and shipbuilding. He favored a government takeover of major industries, but companies like ALCOA would demonstrate that private industry could willingly cooperate with government.

The environment in 1913 after the election of Woodrow Wilson was, for the first time, unfavorable for ALCOA. ALCOA executives and their Pittsburgh East End neighbors had actively opposed Wilson and the Democrats from the start. ALCOA had previously prospered under the Republican administrations. Tariffs were high since Congress's McKinley tariff, but had been reduced under President McKinley from 15 cents a pound to 8 cents as the industry matured. President Wilson reduced the tariff to two cents a pound, and the Wilson tariff had launched a jump in aluminum imports of finished goods. ALCOA's cartel arrangement with foreign producers protected them against primary ingot competition. Prior to the war, Wilson had hoped to break up the ALCOA dominance, and he was hostile to big business in general, but he particularly singled out big Republicans, such as the Mellon family and Pittsburgh Eastsiders.[10]

Arthur Davis headed the company during the many challenges of the Wilson administration and World War I. Wilson was forced, very reluctantly, to back off ALCOA as the war approached. The development of alloys and heat treating never reached that of the Europeans, but ALCOA did prove that it could outproduce its European counterparts. A combination of patriotism and fear of the socialist tendencies of President Wilson made ALCOA one the country's most cooperative industries. By 1914, Davis could speak for the aluminum industry, since ALCOA was the country's only producer.

Davis offered to supply the government all its demands at their own internal price to their finishing operations and related manufacturers. Eventually, ALCOA was sending 90 percent of its production to the government. Wartime production in 1919 reached 130,000 tons versus 60,000 tons in 1914. Still, the Wilson administration was aggressive with price pressure, and ALCOA's increasing profits caught the eye of President Wilson and congressional Democrats. The Federal Trade Commission would investigate the costs of ALCOA and reduced the price from 38 cents a pound to 32 cents a pound. ALCOA continued to maintain high profits because economies of scale were key to large companies like ALCOA. The company did avoid the pressure of the Wilson administration to force unionization of government suppliers, which had been a national effort of the administration.

The real problem for ALCOA would occur after the war, when over-expansion and import dumping would take a toll on ALCOA's patriotism. ALCOA's expansion for the war was far beyond that needed in peacetime. The government had offered little financial aid in this expansion, with the exception of the Tennessee nitrate plant. ALCOA turned down that aid, wanting to keep the Wilson administration out of their operations. The end of the war left ALCOA with overproduction capacity, and Europe was also left with an overcapacity as well. Europe was determined to keep employment high and began to dump aluminum on the American market. The low Wilson tariffs had meant little during the war, but now these tariffs left American labor unprotected. The average aluminum worker in Europe made 11 cents compared to 40 cents in America. As a result, aluminum imports poured in with the tacit support of the American government. By 1920, imports would reach 30,000 tons in a demand-short market.

CHAPTER 8

New Threats and New Markets

THE END OF WORLD WAR I ACTUALLY CREATED a surplus of aluminum and a new push for innovation. Without much government help initially, ALCOA was forced to find and open new markets. In Germany, the surplus aluminum grew as wartime industries were forced to shut down. The Weimar Republic was formed after the defeat of the German Empire in World War I. Coinage in Germany would tell the story of the German aluminum industry. The Republic's early years were a tumultuous period of uprisings, riots, and massive inflation. Germany turned to aluminum coinage, since aluminum was one of the few metals with an excess surplus bringing down its price. The aluminum 50 pfennig was introduced in 1919 to replace the silver half mark that had disappeared from circulation due to hoarding. During 1923, Germany suffered from some of the worst inflation the world had ever seen. In order to keep up with rapidly increasing prices, two new high-denomination aluminum coins were introduced at the beginning of the year, the 200 mark and 500 mark. During all this, Germany continued to produce Zeppelins, expanding into peacetime air transportation.

Britain and France took the opportunity to dump cheap aluminum in America while Russia and Japan saw the necessity of aluminum in war, so they started to build their own aluminum industry in the 1920s. In America, however, things were different. The excess aluminum offered inspiration for ALCOA to look again to new commercial products. One of the prizes of the war was a wealth of German technology seized by the Defense Department, which ALCOA would use to produce its own new alloys. ALCOA would develop a highly decentralized approach to alloy and product development. In addition, the business-friendly Republicans were now leading the nation.

The pro-business Republicans took the White House and Congress in the 1920s. ALCOA would use the decade to expand in all phases of the aluminum business. Congress passed the Fordney-McCumber Tariff Law, which boosted schedules from the average of 27 percent under Wilson's Underwood Tariff of 1912 to an average of 38.5 percent. The Fordney-McCumber Act increased the tariff from two cents to five cents, providing badly needed protection from the dumping of European aluminum. The new tariffs rewarded the sacrifice of ALCOA during the war.

Andrew Mellon had also taken ownership of Gulf Oil, along with other Pittsburgh industries. The 1920s had brought a favorite political environment for vertical integration and expansion after the Wilson administration. Andrew Mellon would lead the charge to change the business environment in Washington. The Woodrow Wilson administration had tried to nationalize large industries and had forced unionization on large companies. It would be Mellon who would lead the growing opposition to the Wilson administration.

Andrew Mellon and Pittsburgh East End opposition to Woodrow Wilson's anti-business practices had taken on an unusual vigor for a political issue in 1919, and Warren Harding was in their camp. Harding was also close to their favorite son and former president, William Taft. Henry Clay Frick, an East Ender and close friend of Andrew Mellon, held a private dinner in 1919 that included George Perkins (United States Steel and J.P. Morgan Company); Dan Hanna (son of Senator Mark Hanna); Ambrose Monell of International Nickel; George Whelan (head of United Cigar); Henry Sinclair (Sinclair Oil); A.A. Sprague (Chicago wholesaler); George Harvey; and Andrew Mellon as well. The dinner was the source of a major fund to defeat the progressive Democrats, and many believed the diners picked Warren Harding as the Republican candidate. Henry Sinclair was assigned the job of building a campaign chest to defeat the League of Nations. The fund was believed to be over three million dollars. Frick was considered the largest contributor, but the list was long and included Charles Schwab, George Westinghouse, Herman H. Westinghouse, John Rockefeller, Andrew Mellon, Walter Chrysler, and Harry Sinclair, amongst many others. In 1920, Harding won in a landslide victory and immediately pushed for higher tariffs.

President Harding would ultimately name two friends of the Pittsburgh East Enders to posts—Andrew Mellon was his Secretary of Treasury and Herbert Hoover (another favorite son of the Pittsburgh capitalists) would be named Secretary of Commerce. Andrew Mellon would become part of President Harding's "poker cabinet." (These poker parties held at "the little green house on K Street" became notorious as the Harding administration's corruption hit the newspapers.) The 1920s saw a very different environment as Andrew Mellon served 11 years as Secretary of Treasury. Of course, his brother Richard Mellon remained on the ALCOA Board. ALCOA was investigated throughout the 1920s for antitrust violations, but nothing ever came of them. However, for the next few decades, the monopoly argument would boil down to one question: Did ALCOA become a monopoly as part of natural business growth, or did it intend to form a monopoly to control the market and prices?

There were other changes after the war as the second generation of ALCOA managers took over. Even prior to the war, ALCOA was moving to the next generation. Captain Hunt, who was a great marketer, had died in 1899. The decade of the 1910s saw the passing of the ALCOA founder, Charles Hall, in 1914. Arthur Davis, co-manager of the Smallman Street plant and Pittsburgh Reduction's first employee, would say at Hall's memorial service: "While it was a great and wonderful thing to invent the process for making aluminum, it was totally different, and actually turned out an infinitely more difficult problem to make aluminum commercially, and a still greater problem to utilize the aluminum when made."[1] The year of 1914 was the eve of war in Europe. The war had not only impacted the aluminum market, but also the management of companies.

The war had created a period of rapid growth on both sides of the Atlantic. Aluminum canteens and goods were used by both sides, and aluminum was truly an international industry. The company was transitioning to an international corporation. The Mellon family increased their share of the company, but unlike other industry financiers, the Mellons worked in partnership with the managing operators. After the war, ALCOA was a company searching for a new infrastructure to support its size, international markets, and competitive challenges. Research managers had to shift from primary production to finding new uses of war surplus aluminum.

The company came under the operational, technical, and marketing leadership of Arthur Davis after the war. Davis had been working under president Richard Mellon, who had concerns about Davis's youth. Richard Mellon focused on company operations while his brother Andrew was involved with finance and international growth. In 1917, Davis was given the top job and he proved to be up to the task in taking ALCOA into the 1920s with a vision for an Aluminum Age. Davis met challenges as great as those that the founders had faced. The challenges would come from many sources. Postwar pull-back of aluminum war products created overcapacity. Henry Ford and others would try to break ALCOA's monopoly of American ingot production by moving into aluminum production. Europe, with its Duralumin alloy, had gained a technical advantage in new products. The 1920s and 1930s would prove as challenging as the early days for ALCOA in terms of labor relations. Labor would become a major input and resource; however, little labor is required in the smelting of ingot. Mining is much more labor intensive than smelting, but still less than most other mining, such as iron ore or coal. ALCOA would have to build towns to support its mining and smelting operations. There would be material competition from stronger and lighter-gauge steel. There would also be new aluminum producer competition for the American market from within, but ALCOA used its advantages to limit competition.

With the armistice in 1919, ALCOA's first challenge was to convert capacity to domestic applications. ALCOA, which had shut down its cookware operations, now needed to restart domestic production. The war had created an extensive network of over 2,000 foundries. These foundries had gained expertise in aluminum casting and had become competitors of ALCOA. With the end of the war they were able to buy cheap foreign ingot or re-melted scrap aluminum. The low quality of scrap aluminum limited it to the sand-casting industry. Still, ALCOA often tried to control the scrap market to give its own foundries an advantage. The Federal Trade Commission investigated these practices and would exonerate ALCOA, but the decision came in an extremely favorable political environment of the 1920s.

The use of the Federal Trade Commission versus the Justice Department was considered political. The Federal Trade Commission was viewed as a weak department with little real power in antitrust cases. In fact, many of the same points would be litigated in the future antitrust case of the 1930s by the Justice Department. Clearly, ALCOA, at least at times, did try to control scrap and scrap prices. In addition, these foundries, such as United Smelting & Aluminum Company, started to move into the fabrication of sheet and wire. ALCOA restricted many independent rolling mills by arrangements requiring scrap buyback at set prices. Meanwhile, ALCOA was aggressively building its own rolling mills.

The 1920s opened with the first sheet aluminum being shipped from its plant in Alcoa, Tennessee. The plant used the Little Tennessee River for power and was the result of prior years of investment. In 1910, ALCOA had created a massive plan for building dams on the watershed of the Little Tennessee River. The Tennessee plant had originally been built to manufacture powdered aluminum explosives, but orders were canceled as the war ended. ALCOA went on with construction of a smelting and a fabricating plant. Aluminum sheet opened up the cookware market by reducing price over cast cookware. A corridor of Wisconsin Aluminum Fabricators had become highly competitive in cookware, using imported sheet where possible. General sheet aluminum manufacturers also arose, using cheap imported aluminum to take on ALCOA sheet. ALCOA countered with advertising and research, as well as better coordination of marketing and product development in sheet usage and cookware.

Competition became very strong in the 1920s as many manufacturers entered the cookware market. This was, of course, direct competition to ALCOA's Wear-Ever division, and it also created indirect competition with independent cookware manufacturers that could buy or roll their own sheet aluminum. For the most part, ALCOA controlled the price of primary aluminum needed by these independent sheet mills, but these independent sheet mills and fabricators still managed to compete. The use of sheet aluminum reduced manufacturing costs over the cast products, generating a new body of cookware fabricators. In 1923, the Club Aluminum Utensil Company introduced Club Aluminum cookware in Chicago. Door-to-door salesmen peddled Club Aluminum, much in the same manner as plastic storage containers, using house parties. Salesmen invited housewives to host a lunch where the salesman would demonstrate the cookware. Club Aluminum was a very creative company and pioneered products like aluminum waffle grills. The new competition forced ALCOA to maintain its product research and keep semi-finished sheet prices down. ALCOA also remained committed to the reduction of primary ingot cost through investment in hydroelectric power.

The production of thinner-gauge aluminum also brought manufacturers, such as metal spinners, into the market to produce a new type of cookware. Metal spinning is actually considered to be one of the oldest methods of forming sheet metal and was used by Paul Revere. In metal spinning, the forming process is accomplished by the use of a spinning lathe. A highly skilled craftsman, called a spinner, forms a flat piece of metal over a specially designed chuck to create the shape. Most of these craftsmen came from Germany, where the technology was a guarded secret among families. Using spinning, a number of new cookware manufacturers entered the market in the United States. Cookware was just one of the new product markets emerging in the 1920s, and much of this new technology was coming from overseas.

ALCOA knew it would have to invest more in research after the war. Francis C. Frary—a brilliant scientist who had achievements in chemistry, chemical engineering and metallurgy—was hired, and started in December 1918. Frary was one of many talented recruits after the war. Dr. Frances C. Frary was born in Minneapolis, Minnesota, on July 9, 1884. He graduated from the School of Chemistry, University of Minnesota in 1905. He then studied at the University of Berlin in 1906–1907. He obtained his Ph.D. degree from the University of Minnesota in 1912, where he was assistant professor of chemistry from 1911 to 1915. He was a research chemist for the Oldbury Chemical Company in Niagara Falls from 1915 to 1917. In 1918, he was commissioned captain in the Ordnance Department and eventually rose to major in the Chemical Warfare Service. Later in 1944, like Hall, he would receive the Perkin Award in Chemistry. There could have been no better replacement for Charles Hall than Frary.

ALCOA Research Center's Art Deco entrance, New Kensington, Pennsylvania (courtesy of Carnegie Library).

Frary would be responsible for improving the Hall Process from 97.75 percent pure aluminum to 99.99 percent pure. Higher purity would help in electrical applications and as a base for alloying. Product research would focus on new heat-treatable alloys to expand the applications of aluminum, which was critically needed to keep ALCOA competitive. Like Hall, Frary would center the research at the New Kensington plant. New Kensington also had a department known as the "Job Shop" for the manufacture of special or custom aluminum products. This incubator at the plant would be the source of many aluminum craftsmen and artists. Frary would bring product craftsmen and artists to expand alloy research. Frary, like Hall, believed in a research function at each plant as well, but would also build a centralized center to help coordinate focus or address corporate-wide issues.

In addition to the new research division centered at New Kensington, all plants would also have a branch research operation. The idea of plant-level decentralized research had been very successful at plants like Buffalo, East St. Louis, and New Kensington, and ALCOA would not change over to full centralization. The idea of a separate centralized research organization had been opposed by Charles Hall, but Davis and Frary designed a combination of centralized function and branch locations at plants. Hall, like most company founders, had struggled as the company grew into a corporation. Hall tended to control the function personally, but Frary and Davis brought a transition. Projects had to be focused and strategic to the company objectives. Through the 1920s, scientists, artists, craftsmen, and engineers were hired in an effort to catch up with the Europeans using a more centralized approach. ALCOA had much ground to make up after the war in terms of research and development, and strategic research goals would help. The Europeans led the industry in heat-treatable high-strength aluminum alloys, and ALCOA started there.

With the captured technology of the Germans, ALCOA finished its development of its version of Duralumin known as 17S in 1922. The development of this alloy would renew interest in the American dirigible program. The earlier program of the Navy was a failure, as several dirigibles crashed in the mid–1920s. The new 17S alloy allowed for ALCOA to develop its rolling mills to produce girders, but the Navy lost interest in the program, so ALCOA then turned to civilian applications of the new alloy. One product that they developed was aluminum furniture, which had first been sold in Germany as early as 1906 using Duralumin. The Austrian Postal Savings Bank had aluminum chairs in 1906 made with the first Duralumin alloys, and aluminum chairs also became popular in France by 1910. ALCOA, with its 17S alloy, finally had the product strength for chair manufacturing in the 1920s. Russia, however, was a step ahead of ALCOA in civilian applications at that time.

The fuselage of a shot-down Junkers D.I fighter was delivered to the Russian Research Institute (Central Aerohydrodynamics Institute) in 1918. A separate Material Testing Division group was organized in order to study the composition of the airplane's metal covering. Russia marshaled its resources to break the code of Duralumin production. Like Germany, Russia looked at aluminum development as a national and strategic research function of government. The Russian researchers did not just determine the formula of Duralumin, but managed to develop a stronger alloy modification, able to compete with foreign developments. The results of their work were sent to the brass and copper-rolling plant of Kolchougin Co. and the Leningrad plant Krasny Vyborzhets. The first to master the production of this Russian know-how were the metallurgists of the Kolchougin plant. In late 1922, the plant started production of "kolchougaluminium"—the first Soviet high-strength alloy, which was actually stronger than Duralumin. Work was started to create a competitor to the Junkers aircraft, and the Soviet airplane AN-2 came out in 1924. The 1920s saw aluminum expand into new Russian domestic markets with the advent of strong alloys. The Russian and American versions of Duralumin opened the world to a strong aluminum that could compete with steel.

Duralumin and new cast alloys offered strength that had originally been lacking with earlier versions of aluminum. The lack of strength had limited aluminum to being used in bicycles, and it certainly ruled out motorcycles. A historic year for the motorcycle industry and aluminum was 1922. Carlo Borrani brought the Rudge-Witworth International Patent production of their cast aluminum motorcycle rims. Borrani set new standards for the motorcycle road and racing industry with cast aluminum. In the late 1920s, aluminum fuel tanks and engine parts were added. This new product introduced to the market provided unrivaled strength, lightness, and precision to motorcycle racing. The quality of Borrani's aluminum wire wheels came from Borrani's choice of light-alloy aluminum, performing drilling tailored to the types of wheel hubs, and finishing every rim by hand.

Duralumin offered a new material for domestic furniture builders as well, and ALCOA moved into high-quality furniture production. "In 1924, ALCOA furnished Mellon Bank headquarters in Pittsburgh, and aluminum furniture was soon employed in the Free Library of Philadelphia, the Waldorf-Astoria Hotel, and the New York Insurance Company. In the late 1920s, the ALCOA factory in Buffalo produced aluminum chairs for sale in European as well as in home markets. Each weighing less than eight lbs., these chairs were sandblasted, primed, spray-varnished, and then coated twice with enamel color. The standard colors were walnut, mahogany, and oak."[2] Europe was not far behind in this high-end market, especially in France, where the metal was still

A passenger Zeppelin in 1908 (Library of Congress).

considered semi-precious. France and Germany were quick to bring in artists for designs; and, in 1929, there were a number of French and German aluminum furniture producers who favored the luxury market. By 1930, in Germany, aluminum chairs were booming business. Ludwig Mies van der Rohe designed his famous *Chair and Stool* in 1929, originally created to furnish his German Pavilion at the International Exhibition in Barcelona. While it had come to epitomize modern design, it was not fully aluminum, but did contain aluminum rivets. Mies had dreamed of using aluminum, although he felt alloys of the time still lacked the strength, but through design and alloying, he improved his chairs. Mies van der Rohe designed a chair to serve as seating for the king and queen of Spain, and stools were intended to accommodate their attendants.

ALCOA pushed hard to develop new strong alloys for furniture, too. A plant in Buffalo, which had cast Pierce-Arrow car bodies in the 1910s, was converted to manufacture furniture. The market demand took years to develop, as the growth came slowly in America because the connection between aluminum and art never developed as in Europe. The 1920s were a period of major transition in European interior design, marked by experimentation with new materials and modernized design. Americans were slow to accept modernism, and the American high-end aluminum furniture industry remained weak. ALCOA was successful by using traditional designs that appealed to conservative American tastes. In 1929, ALCOA manufactured over 35,000 chairs, and they continually added new designs and built a mystique about aluminum furniture of its own. Many Hollywood mansions, such as that of Marlene Dietrich, were adorned with aluminum furniture. The acceptance of Hollywood opened a booming market. By the 1930s, it was a market of considerable size, but ALCOA opted out of selling the furniture and sold its furniture division, preferring to supply semi-finished aluminum. Still, there would be a strong market for ALCOA tubing, while developing individual manufacturers of finished goods. The strategy was basic to ALCOA to develop a finished-goods industry and then back off to be a supplier versus a manufacturer. ALCOA used its research center to help develop and support emerging markets for aluminum.

Another specialty item developed at the ALCOA Buffalo "research center" or "skunk works" was the musical bass, although the idea of aluminum musical instruments was not totally new. In 1891, a Mr. Alfred Springer was awarded a U.S. patent on an aluminum violin. Springer's violin had a body that was made of aluminum sheet. The top, back, and ribs were attached with very small rivet-like pins that were flush with the sides. The neck and fittings were wooden and a traditional spruce sound post was used. In 1894, Neil Merrill founded the Aluminum Musical Instrument Company to produce violins, mandolins, guitars, and cellos. Customers found the feel and look to be a major problem, however, and the company was bankrupt by 1898. The ALCOA Buffalo plant produced 500 aluminum basses that sold for around $250 for a New York musical supply house in 1928. ALCOA then worked with the Interlochen School of Arts in Ann Arbor, Michigan, to design a cheaper model which sold for $50. ALCOA even produced a patented faux wood surface and a gold painted surface. The market never fully developed, though, partially because the sound was different from wood, and the surfaces were affected by temperature. However, before the Buffalo plant closed 1934, it produced 435 aluminum violins for a Dr. Joseph Maddy of Interlocken School of Arts, which continued to sell into the 1970s.

One area where ALCOA had a worldwide advantage was with precision-cast automotive pistons. ALCOA expanded the process to make diverse parts like aircraft cylinder heads and washing machine agitators. ALCOA also developed a line of alloys with silicon to allow the liquid aluminum to flow into intricate designs and developed a new heat-treatable forging alloy, 25S (today's 2025), to use in forged propellers. In 1926, ALCOA improved on this alloy to make forged aircraft engine crankcases and started an experimental line of forged aluminum pistons. A new aluminum engine, the Ryan NYP, would be highlighted in Lindbergh's successful transatlantic flight in 1927. Lindbergh's plane, *Spirit of St. Louis,* also used aluminum sheathing. Still, aluminum was not fully applied in aircraft skins due to corrosion issues, as saltwater environments caused cracks to form in alloy aluminum.

ALCOA and the federal government cooperated to develop a material known as Alclad, which consisted of an aluminum alloy bonded to pure aluminum. This was a strong heat-treated aluminum alloy core with a smooth surface of pure aluminum integrally bonded to the core. The pure aluminum coating supplied superior resistance to corrosion and cracking. The corrosion resistance would be key in the use of aluminum in aircraft skins, which faced aggressive salt environments. The first aircraft to be constructed from Alclad was the all-metal Navy airship ZMC-2, built in 1927. In the 1930s, the research on Alclad started to pay large dividends as it led to all-aluminum planes.

One setback for ALCOA's product development was in sheet products. Prior to the war, aluminum sheet had made inroads into the auto market. In 1920, there were three independent aluminum sheet-rolling companies, and some were beating ALCOA on price and quality by using imported aluminum. The auto recession of 1921–22 left ALCOA with excess capacity, while others expanded. ALCOA had invested heavily in the sheet market and expanded the New Kensington plant for production, but it was still not competitive. In particular, Brush Machine Company had built a powerful sheet-rolling mill in 1919. The auto industry sheet demand never developed after the recession because, ironically, aluminum killed steel, and Pittsburgh Reduction became cheaper and better for making auto body parts. ALCOA preferred to expand in cast parts versus sheet.

Expanding in the cast aluminum industry, ALCOA purchased Acme Die Casting Corporation in 1926. Casting was one of the areas that ALCOA broke with in its strategy of staying out of the finished goods market. These new alloys were aluminum zinc, which allowed liquid metal to be pressure flowed into a metal mold. In 1922, over 70 percent of its die castings were going to make Hoover vacuum cleaner parts. By the late 1920s, ALCOA was making die-cast auto parts such as gas caps. In 1930, Sears Roebuck used cast aluminum in its Imperial Kenmore vacuum cleaner.

Another ALCOA innovation was the collapsible tube for the production of toothpaste tubes in 1921. These tubes became extremely popular, and a plant was built in Edgewater, New Jersey, in order to produce the tubes. ALCOA also invented a number of protective aluminum coating systems. The company developed a process of rolled-on sealant, which sold as Mason jar caps, and created an ALCOA subsidiary called the Aluminum Seal Company. In 1926, the company produced over 25 million jar caps, and ALCOA looked to strengthen its centralized research with this product success.

A research center was proposed to replace the "skunk works" at New

Henry Hornbostel's aluminum spire on Pittsburgh's Smithfield Church with ALCOA Building in background, 1949 (courtesy of Carnegie Library).

Kensington. The ALCOA Research Laboratory was built at New Kensington and the Allegheny River Valley in 1929. The building would be a symbol of ALCOA's innovations. With its design showing influences of neoclassicism and the Art Deco style (also popular with aluminum designers), the plans were executed by renowned Pittsburgh architect Henry Hornbostel. In 1926, Hornbostel had built an aluminum spire to top the Smithfield United Church in Pittsburgh. The ALCOA research building itself was a tribute to aluminum in architecture and building. The multi-light windows of the first and second floors were separated by a band of ornate aluminum panels. An intricately detailed aluminum railing served as a wall and it extended around the flat roof of the main building. A stone stairway led to the entrance, which featured ornate aluminum gates, and the double doors leading to the interior entrance hall containing panels with floral and shamrock motifs of hand-wrought aluminum. The entrance hall had marble floors, and aluminum stair railings, light fixtures, baseboards, heating grates, doorjambs, and doorknobs were found in abundance throughout the main building. There are also crafted aluminum chairs and desks, and in the basement, there were three ornate aluminum bookcases reportedly made for the Hunt family.

The American production of aluminum foil from sheet became a specialty of Reynolds Metals, the company that over took ALCOA in the late 1920s, as ALCOA never entirely committed to the full development of this market. The Reynolds Metals Company was founded in 1919 as the U.S. Foil Company in Louisville, Kentucky, by Richard S. Reynolds, Sr., a nephew of tobacco king R.J. Reynolds. The company started by producing lead foil and tin foil wrappers for cigarettes, food, and candy decades before. In 1926, the company first began production of aluminum foil, resulting in an expansion into aluminum products including foil bottle labels, foil bags, and insulation paper. Aluminum packaging even led Reynolds to purchase Eskimo Pies, which in 1924 was a major user of foil. Still, aluminum foil did not overtake tin foil until after World War II.

Foil rolling was a difficult process. The process required rolling and re-rolling sheets with a mandatory step of heating in between. Aluminum foil has a shiny side and a dull matte side; the shiny side is produced when the aluminum is rolled during the final pass. It is difficult to produce rollers with a gap fine enough for the final pass; therefore, two sheets are rolled at the same time, doubling the thickness of the gauge at entry to the rollers. When the sheets are later separated, the inside surface is dull, and the outside surface is shiny. The shiny surface is often preferred where heat needs to be reflected, as in insulation. Final thickness in the 1920s was 0.006 inches thick or 2 sheets of paper thick. Today, foil can be made of thicknesses down to 0.0015 inches thick. Foil fit well into ALCOA's rolling of narrow-gauge sheet, which offered bigger markets in the 1920s.

ALCOA preferred to develop higher tonnage markets in the 1920s. ALCOA did have a large market in narrow sheet for cookware, but it had serious competition in wide sheet. By 1925, the future for wide aluminum sheet appeared to be in aircraft, rail cars, and trucks. ALCOA decided to control primary aluminum ingot going to these competitive independent rolling companies. These borderline monopolistic practices would become the root of legal problems for ALCOA. Brush Foundry filed legal lawsuits claiming they were being restricted entry into the market by ALCOA. ALCOA was the only ingot producer in the United States, allowing it some control. ALCOA would win the court challenges, but the market still lagged behind expectations. Still, in 1929, ALCOA gambled on a three-million-dollar sheet mill at Massena for a future market that did not fully develop until the late 1930s. The company continued to keep pressuring its researchers to find new markets for wider sheet.

Even the company's friends in government couldn't prevent antitrust lawsuits from coming against ALCOA in the 1920s. Clearly, ALCOA had used forward integration to develop new markets for its product. ALCOA had built fabricating plants to produce new products and became a monopoly in its quest to develop uses for aluminum and control primary aluminum ingot production. The fact was, ALCOA was indeed a monopoly, legal or not, in complete control of the American aluminum market. However, the main question was, had this happened as a price control strategy or as a strategy of product development? ALCOA's lawyers would argue: "ALCOA initially entered fabricating primarily because of lack of interest in aluminum by existing metal fabricators. In end products, however, ALCOA followed a general policy of persuading existing end-product manufacturers to produce new aluminum products, and only after indifferent or no success did ALCOA produce end products, and then only temporarily and on a small scale."[3] The argument would stand, at least in the business-friendly 1920s, and ALCOA would win all its lawsuits through the decade.

On an international level, ALCOA was viewed as a cartel. Through the 1920s, ALCOA expanded in Europe by purchasing foundries in Germany, Spain, France, and England. Additionally, ALCOA expanded its purchases of bauxite mines around the world. ALCOA also moved into European smelting by purchasing Norsk Aluminum of Norway. ALCOA built and purchased hydroelectric plants in Spain and France to produce primary aluminum ingot. ALCOA reached 15,000 tons of production in Europe, compared to 120,000 tons total produced by Europe, and compared to ALCOA's American production of 83,000 tons. In 1928, ALCOA controlled over half of the world's primary ingot production, and the international expansion didn't go unnoticed or unchecked. The 1920s saw a rise of nationalism in Europe. In England, there were "Buy-British" campaigns. European nations were also increasing tariffs on incoming aluminum products.

ALCOA looked to form a European-based company that, while "independent," would not become a competitor. Another reason to form this new company was because of the restriction by the Sherman Antitrust Act preventing

American companies from participating in foreign cartels. European-style cartels would also help ALCOA control world bauxite prices. ALCOA, realizing the pressure from national governments, decided to form a cartel with its international competitors. Through this cartel, markets were divided up and price levels were maintained. Clearly, ALCOA was a growing monopoly and cartel, not from natural business growth, but rather with the intention to exercise control over the marketplace. With high prices supported by the cartel and domestic price control of primary ingot, cost reduction in smelting and primary production went straight to the bottom line. ALCOA's primary focus became process cost cutting in its smelting operations.

In the 1920s, James Duke, the tobacco tycoon, would enter the hydroelectric power industry and ultimately make Canada a competitive aluminum producer. Duke offered another way to reduce costs through the use of hydroelectric power. James Duke and Arthur Davis teamed up to create the embryo that would become the second largest aluminum company in the world. The Saguenay region of Canada is a center of vast resources for water power as the Saguenay River is the largest tributary of the St. Lawrence River. The Saguenay River is fed by a powerful drainage flow from Lac Sainte-Jean through two deep gorges. The water drops some 300 feet through a deep gorge and runs 9 miles to form the headwaters of the Saguenay River, making it ideal for hydroelectric power plants. In 1912, Canada had tried and failed to harness the power, but James Duke was able to turn the project into success.

James Duke (1856–1925) is best known as the founder of Duke Power and Duke University, but he had a very diverse career. Duke had made his fortune in building his North Carolina industry into a huge trust known as the American Tobacco Company. In 1906, the American Tobacco Company was found guilty of antitrust violations and was ordered to be split into four separate companies: American Tobacco Company, Liggett and Myers, R.J. Reynolds, and the P. Lorillard Company. Afterward, Duke moved into the textile industry in the Carolinas and started a textile mill in 1892, which led to the application of hydroelectric power generation to reduce costs.

Duke organized the American Development Company to acquire land and water rights on the Catawba River. In 1904, he established the Catawba Power Company; and the following year, he and his brother founded the Southern Power Company, which became known as Duke Power, the precursor to the Duke Energy conglomerate. The company supplied electrical power to Duke's textile factory and, within two decades, also supplied power to the Piedmont region of North and South Carolina. Duke became a hydroelectric enthusiast with other capitalists such as Henry Ford and Thomas Edison, and it was that enthusiasm that drew him to the Saguenay region of Canada.

In 1912, Duke embarked on a project for a power plant in the Saguenay region. He moved to start construction in 1913, but Duke realized he was ahead of the demand for power in the area and put the project on hold. Duke, in 1920, teamed up with the Price brothers of Canada to supply power to their proposed newsprint paper mill. Duke started construction of the Isle Maligne hydroelectric plant, which would become the world's largest in 1925. But even as the Duke-Price project approached full operation, there was a shortage of customers for this super plant.

Duke had started to look at aluminum manufacturers as customers for his power surplus. One of these manufacturers was George Haskell, owner of a major aluminum fabricating plant and sheet mill, but the plant lacked primary ingot production. At the same time, ALCOA was falling short in its ability to supply aluminum to these independent sheet mills. In 1925, ALCOA was 55 million pounds short, forcing the companies to buy from Germany. ALCOA had been involved in the early failure of Canada to produce power in the Saguenay region, but Duke's success had generated new interest. ALCOA's needs and Duke's shortage made for the perfect marriage. In 1925, ALCOA merged with Duke-Price, with ALCOA being in control of the "new" company. The merger knocked out Haskell and his effort to produce primary aluminum; and, in return, Haskell would file antitrust suits in 1925, 1926, 1932, and 1937. While these suits would result in many court battles, Haskell maintained that it was purely business. He later testified that ALCOA was never trying to intentionally injure or put anyone out of business. Eventually, ALCOA would win, but then lost the appeal, which forced a settlement with Haskell; still, the merger would make ALCOA the world's largest aluminum company and give it monopolistic control.

ALCOA began construction of a Canadian smelting plant in 1925. Because of poor land around the Canadian Isle Maligne hydroelectric plant, the actual smelting plant was built down river on some acquired farmland. The site, known as Arvida, would be an iconic example of a new type of paternalism. ALCOA was required to build rail connections as well as a new town. Arvida will be discussed in a later chapter. In addition, ALCOA had to purchase extensive new supplies of bauxite, and started to buy deposits in the British colony of Guyana. The end result was stiff political pressure from England over the ALCOA world monopoly.

CHAPTER 9

Birth of Alcan and the Rise of an International Industry

THE YEAR 1928 STANDS OUT IN THE HISTORY of the aluminum, as world production had reached 250,000 tons, with Europe accounting for 130,000 tons. In North America, all production was controlled by ALCOA, with 90,000 tons in the United States and the other 42,000 tons produced by its Canadian subsidiary. The market for aluminum was centered in the United States; therefore, Arthur Davis and Richard Mellon believed the focus needed to be there. ALCOA also wanted to expand fabricating plants in the United States, and had hoped that the spinoff would focus on ingot production. In hindsight of many decades, the split seems questionable because it would ultimately, in the long run, create ALCOA's biggest competitor.

ALCOA favored smelting in Canada and had invested heavily. Canada was a logical location for aluminum smelting with its vast water power resources. Today, Canada is the world's second largest producer of hydroelectricity after China, which has more than double Canada's level. In 1928, Canada stood alone in total hydroelectric power, which was one of the most efficient sources of energy. Hydropower stations can convert more than 90 percent of the available energy in the river into electricity, while most conventional fossil fuel plants were less than 25 percent efficient at the time.

In 1928, the idea of splitting ALCOA into a U.S. company and an international one made sense. However, there was a downside to giving up the world's most efficient smelting plants. An international company would not seem to be a major competitor at the time. The Canadian-based international company would have an excess in ingot production and a lack of fabricating plants and markets. The plan was for the Canadian company to supply the fabricating plants in the United States. The disadvantage for ALCOA was the loss of the Canadian smelting operations, which would become the foundation of a future competitor, Alcan. The decision seemed to have been made with little internal opposition at ALCOA, or more likely, no one wanted to oppose Davis.

The record is silent on the Mellon position. Andrew Mellon was still serving as Secretary of the Treasury and was somewhat out of the loop. Richard Mellon maintained stock control, but had always been easygoing in his approach and easily dominated by strong personalities such as Davis. Mellon biographer noted: "As long as Davis remained chairman, Mellon constituted a voice, little more, another opinion worth considering."[1] Davis had taken over as president of ALCOA in 1910 when he replaced Richard Mellon; and, in the end, this was Davis's project and he was a strong Napoleonic leader.

Also in 1928, ALCOA transferred most of its foreign assets into a newly formed British company, Aluminium Limited. The company would be managed by former ALCOA management; but on paper, Aluminium Limited would be a separate identity. Davis planned to control Aluminium Limited through a type of interlocking management arrangements. The younger brother of A.V. Davis, Edward K. Davis, became president of the new Aluminium Limited. The selection of his brother assured tight control; Edward was a seasoned executive and experienced in the aluminum industry.

Edward Davis was 13 years younger than his brother with 25 years of aluminum manufacturing experience. Edward had graduated from Harvard in 1903, but started his career at the bottom at ALCOA. He began at the Canadian smelter at Shawinigan, Quebec, making 75 cents a day with an additional 12 cents a day for living expenses. Within a year, he was transferred to the smelter at Massena, New York, where he worked for 12 months. Edward would work at several fabricating plants before going to New Kensington, where he was the assistant to the rolling mill superintendent. He rose to mill superintendent; and, in 1909, he was moved into sales at the Pittsburgh headquarters. Just before 1914, he became the head of international sales and marketing. His love of international sales made him the ideal choice to be president of Aluminium Ltd. (Alcan Company). Roy Hunt, son of founder Captain Hunt, became president of the new ALCOA Company in the United States, which ultimately resolved the competition that had arisen between the two.

Aluminium Limited ALCOA stockholders were given stock shares in the new company, making them owners of both companies. Major stockholders of the two companies continued to be identical until 1951, when a U.S. District Court compelled them to sell their stock in one company or the other, in order to remove the appearance of collusive action. Davis had hoped that this paper arrangement would protect them against antitrust suits. Aluminium Limited also had all European operations, and this was part of a monopolistic cartel that incorporated sharing stockholders and resources, including product research.

While most of these foreign fabricating plants were supplied ALCOA primary ingot, the company continued to expand by building European plants and employing Europeans. Much like the Japanese car transplant factories of the 1980s in America, these physical plants reduced the criticisms of nationalists in Europe and provided a tactic by which to dodge high tariffs. The extent of world market control by the late 1920s allowed ALCOA to reduce, if not totally prevent, dumping of low-price aluminum ingot in America.

Another reason for splitting aluminum companies into two was to get control of new bauxite deposits in the British colony of Guyana. In 1917, mines were being developed in Guyana in hopes of backing the allies' need for aluminum. By the early 1920s, it was obvious to ALCOA that the domestic supply of bauxite was far short of what would be needed for future production. ALCOA had been buying mines there quietly. In Britain, there was a rise of empire-building and nationalism. Many in Britain had become jealous of the control of the world's raw materials by United States. Guyana was not only a large deposit of bauxite, it was Britain's only known source at the time. ALCOA had been buying land and promoting the use of this ore for its Canadian operations with the goal of placating British politicians.

Many politicians pressed for the Guyana mines to be nationalized and ceded to British Aluminium. The situation in 1928 can be summarized as: "The British government sought to prevent further control of Guyana's bauxite falling into ALCOA's hands and to secure an Empire source of bauxite for the British aluminum industry, which at the time relied on continental sources of supply which were considered particularly uncertain in the event of war."[2] The formation, or split, of the company into ALCOA and Aluminium Limited allowed for a "British" company control. Initially, Aluminium Limited was dependent upon ALCOA, as they purchased alumina from ALCOA to run its smelters. Also, much of the Canadian ingot was processed into sheet and bar in ALCOA mills, done on a toll basis. Edward Davis set a plan to develop their own rolling mills and achieved independence by 1935 in order to satisfy local politicians.

Given hindsight of history, the split appears to have been done more for personal reasons rather than for business. In 1928, there was little to fear from the government in terms of antitrust. Although there were inquiries by the Justice Department, ALCOA was cleared. Also in 1928, the Federal Trade Commission investigation was still ongoing, but the outcome was believed to be that ALCOA would win. The Coolidge administration was known for its laissez-faire and small-government approach, and Andrew Mellon was Secretary of the Treasury in the administration. On the international scene, nationalism problems in Europe could have easily been taken care of without spinning off the Canadian super smelters. Arthur Davis, in a June 1939 antitrust hearing before Congress, testified about the personal problem, while stating it was not the prime reason.[3]

However, the issue was that Arthur Davis was 61 years old at the time of the split, and he wanted to settle corporate succession. There were two equally strong candidates for future president—Edward Davis, Arthur's brother, and Roy Hunt, son of the founder. Roy Hunt was the Mellon favorite, and Davis would have problems moving his younger brother. The split allowed both to become president of a company. Edward Davis would be president of Alcan while Roy Hunt was the president of ALCOA. The Mellons may have agreed because they would be in control of both companies and because they were unaware of what would come for the business's future. The company would still have monopolistic control at the time of the split in 1928.

There was another pressing reason for Davis to make the split before 1929. Arthur Davis had been extremely close to the founder Charles Hall, and since Hall had never married, he made Davis co-executor of Hall's estate with Oberlin College. More importantly, Davis had the right to vote the common stock shares in Hall's estate until 1929. Roy Hunt also had considerable stock; and he preferred, and voted along with Davis, to block Mellon's control. Davis had effective control of the board of directors and, in 1925, he had been involved in some efforts to block the Mellons. He realized things would change in 1929. The Mellon family probably controlled 30 percent of the stock in 1928, but with the lapse of the Hall trusteeship in 1929, they would have effective control. If Davis wanted to assure the split on his terms, he needed to act prior to 1929. Roy Hunt, like Edward Davis, was well qualified to be president of the company.

Roy Arthur Hunt was born on August 3, 1881, in Nashua, New Hampshire. He was the only child of Alfred E. Hunt and inherited Alfred Hunt's stock in the company. Soon after his birth, the family moved to Pittsburgh, where his father Alfred founded the Pittsburgh Reduction Company. Roy Hunt graduated from Yale and began his career at the New Kensington plant. By 1907, when the company changed its name to ALCOA, Roy had worked his way up to assistant superintendent. In 1914, Roy joined ALCOA's board of directors and served as general superintendent of the company's fabricating plants, where he helped accelerate production of aluminum for the war. Like his father, he had the rare combination of technical and people skills. In 1918, Roy was promoted to vice-president of both the fabricating and the smelting plants. Then, in 1928, he was elected president of ALCOA and served in that capacity until 1951 when he was named chairman of the Executive Committee, a position he held until 1963. Roy remained a member of the ALCOA Board of Directors until his death on October 21, 1966. Roy Hunt managed in the same style as his father. Roy loved an informal and family-oriented business that was low on structure and organization. Davis wanted Roy and his brother Edward to take a leadership position; but after 1929, the Mellons would have control. The 1928 split allowed Roy and Edward Davis to take major leadership positions. In the end, Roy Hunt would become one of ALCOA's finest leaders.

Upon retiring, Roy Hunt noted his approach to management, saying: "In my Credo there are 5 allegiances that we have. I put them in order of their importance.

(1) To Almighty God: broadest mainstay of spiritual destiny.
(2) To our country: no need to elaborate this one.
(3) To one's self, including one's family.
(4) To the organization for which you work—and also in which you have an interest.
(5) To charitable, civic, and community organizations, which, as one grows older, take more and more time.

To all of you I commend these five. They constitute my own philosophy."[4]

Another challenge to ALCOA's market in 1928 was that of the new light metal, magnesium. Magnesium is the lightest structural metal in the world, approximately 34 percent lighter by volume than aluminum, and it is the 8th most abundant element in the Earth's crust. Aside from its light weight, a few other advantages that magnesium offers include excellent fatigue resistance, denting resistance, and the highest known vibration-damping capacity of any structural metal. Magnesium machines better, but the chips are highly flammable. Reviews of ALCOA's participation and market control suggest that ALCOA possibly retarded the development of magnesium to maintain its monopoly of light metal.[5] Certainly, in the 1910s, ALCOA realized a potential threat and actively entered into magnesium production; and eventually, that concern led to the purchase of American Magnesium Corporation in 1919.

Using the rights to American Magnesium, ALCOA became a producer of magnesium at its Niagara Falls plant. The American Magnesium process for magnesium was like aluminum, an electrolytic process. Magnesium oxide and fluoride were used for the molten bath. ALCOA moved rapidly into wire, sheet, tubing, and castings using its aluminum mills, and formed a subsidiary company, American Magnesium Corporation. The main competing process was the patented process of Dow Chemical. The Dow process was considerably cheaper and used magnesium chloride mined in Michigan or from salt water, then used simple electrolysis to make magnesium from a liquid solution (with no heating of the bath). The Dow process was considerably cheaper. While ALCOA became the largest magnesium producer in 1927, its ingot cost was much higher than Dow, forcing ALCOA to purchase all magnesium ingot for its mills from Dow Chemical.

In 1931, ALCOA entered an agreement with I.G. Farbenindustrie of Germany to use their patented process. Germany had been the leader in magnesium manufacture. The 50–50 joint venture of the ALCOA and I.G. Farbenindustrie was called the Magnesium Development Corporation. By 1933, ALCOA and Dow were forced into a loose agreement, and the arrangement would eventually lead a to conspiracy case by the Department of Justice. The case would cause a public debate between Dow and ALCOA, with Dow believing it had no choice but to enter the arrangement due to ALCOA's market control. But by the start of World War II, the case would be settled with no charges against either party. In the end, the magnesium market boomed but never really challenged aluminum as the light metal leader. Today, magnesium alloy consumption is less than one million tons per year, compared with 50 million tons of aluminum alloys. Its use has been historically limited by its tendency to corrode, the high temperature creep, and flammability properties. Still, magnesium is the third most commonly used structural metal following iron and aluminum.

The only challenge to ALCOA's American ingot monopoly would actually be prevented by the federal government itself. That challenge would come from industrial titan Henry Ford. From 1901 to 1929, many automakers had produced aluminum body cars in limited quantities. Henry Ford produced his first experimental aluminum body car in 1909, when the Model T was made of wood. These handmade aluminum bodies were made in the hundreds, but Ford soon went back to wood. However, in the early 1920s, Ford produced a factory-built aluminum car. Introduced in late 1922, the new 4-door sedan was added to the Ford line and was priced at $725, by far the most expensive catalogued Model T, which was known as the Fordor. The price was more than double the cost of the non-aluminum Model T Touring Car that was only $348. The Fordor retained its distinction as the most expensive sedan in subsequent years but, nevertheless, it sold in serious numbers: 144,444 in 1923, 84,733 in 1924, and 81,050 in 1925.

Through the 1910s, Henry Ford had been increasing his use of aluminum in the Model T as well. Then, in the 1920s, Ford planned to further increase aluminum in his Model T, reduce its weight, and make it more fuel-efficient prior to the war. After the war, it was the price of aluminum that held Ford back from his plans. Ford, a believer in hydroelectric power, began to consider a move toward manufacturing aluminum in 1918, believing he could substantially reduce the price of aluminum. Ford and his friend Thomas Edison searched the country for a potential plant site. Ford and Edison also had a passion for the use of hydroelectric and hydromechanical power. Many of their famous camping trips centered on the study of potential water power. Ford later used water power for his home and for some of his smaller factories. In the 1920s, he was searching for a hydroelectric source to produce aluminum cars.

The center of Henry Ford's interest was a 30-mile stretch of rapids on the Tennessee River, known as Muscle Shoals (the nearest town) in northwest Alabama. Ford and Edison had camped on the site and explored the Muscle Shoals area in search of its potential to be "the Niagara of the South." Thomas Edison said of Muscle Shoals: "The possibilities here are greater than all the gold in the world."[6] For perspective, Muscle Shoals is about 60 miles from Huntsville, Alabama, 120 miles from Memphis and Nashville, Tennessee, and 200 miles from Tuskegee, Alabama. ALCOA had been buying land in the area since 1910, but saw more potential in its Canadian hydroelectric resources. During World War I, the government had built a dam there (the Wilson Dam) and had planned to use the hydroelectric plant to produce nitrates for the military. Today it is part of the Tennessee Valley Authority of great dams. This part of the Tennessee River Valley was, in 1922, home to over four million people, most of whom were farmers in poverty. This area of the river, with its swift rapids and falls, had been an early block to navigation of the Tennessee River in the 1800s. The surrounding area was made up of small, poor cotton fields. As early as 1820, many sought federal aid to develop canals on this portion of the river for improved navigation for the cotton growers.

Ford's plan was for a massive utopia of a city and surrounding farm district. The 37-mile stretch of river around Muscle Shoals dropped 134 feet in elevation, which offered even more potential than Niagara Falls. In 1916, with World War I looming, the Wilson administration became interested in the potential of Muscle Shoals to generate electricity and produce nitrate fertilizer for cotton growers and nitrate for explosives and munitions. Wilson and his progressive Democrats wanted this to be a government project. National defense hawks and "King Cotton" congressmen were able to pass the project for electrical production at Muscle Shoals. The government ultimately spent over $130 million on building a dam but never completed the project. It was considered a white elephant and quite a big embarrassment for the government.

At the end of World War I, the government lost interest in the completion of the project, and Congress denied further funding. The government had already spent more than $105 million on the stalled project so far, and was paying over $300,000 a year just to maintain idle installations. The assets lay wasting away with no hope of funding in 1922. Ford saw an opportunity to move in and finish the project, but he dreamt on a much larger scale. Ford envisioned a 75-mile-long city that he would call "the Detroit of the South" and employ over a million people. There was to be electric passenger and freight service to Memphis, Nashville, Chattanooga, Atlanta, and Birmingham. Just as big was

Ford's dream for the farmers of the Tennessee Valley. Ford planned to manufacture all three components of commercial fertilizer—nitrates, phosphates, and potash. Nitrates would be made from electricity and coke; phosphates would be quarried and crushed; and potash would be gained from Georgia shale. Ford believed he could produce these at the lowest cost in the world. With available fertilizer, Ford even predicted the poor cotton growers in the surrounding area would get a half a bale more per acre. For Ford, it was all part of his industrial-agricultural vision for the area.

However, Ford realized that he would need other resources and infrastructure to support such a massive industrial city. Ford believed he could produce aluminum far cheaper than other world producers, and aluminum would be so cheap that he envisioned building aluminum farm machinery. He planned for huge power usage at Muscle Shoals with a massive aluminum plant that would function year-round. For his electrical city, additional power would be necessary in seasonal low flow times on the river. Ford had purchased extensive coal lands in Kentucky about 200 miles from Muscle Shoals as alternative energy, even before he had official approval to purchase the site from the government. Ford's plan was to maximize efficiency by burning coal at the Kentucky mines, generating power to send to Muscle Shoals versus shipping coal to Muscle Shoals and wasting transportation. Ford also started to buy bauxite mines in surrounding states.

Ford saw this electrical city as free of smog and free of waste dumps. To keep his electric metropolis's air clean, he proposed that aluminum cars in the area should run off grain alcohol made in the surrounding farms, and aluminum trains would do the long hauling of materials and people. Waste materials, such as ash, would be utilized in building materials. Ford put the cost of the project at $45 million; and although he had the money to do it, he did not have ownership of the land, yet his grand view gained plenty of popular support from farmers and unions. Ford's project received support from the National Grange, the Farmer's Union, the Federal Farm Federation, and the American Federation of Labor (AFL). A national organization called "Give Henry Ford an Opportunity Club" solicited $1 to $25 donations. One massive rally in Mobile, Alabama, drew over 5,000 supporters.[7] Ford proposed he would pay the government $5 million in cash and $1.5 million a year in rental fees. Ford's estimated fortune at the time was probably in excess of $1 billion, so he clearly had the money for such a grandiose plan.

Along with many supporters, Ford also had powerful opponents as well. First were the progressive Democrats who wanted this to be a government project. The progressive movement of the time believed in government ownership of power generation, and the opposition to Ford's plan became formidable as enemies united. Ford's own opinions always seemed to hurt him in the political arena, as his anti–Semitic views had made the papers across the country, and many had forged political forces against him. His support of black workers had made him unpopular in the South, and the idea of improving the economic lot in Muscle Shoals area was not popular amongst all white residents at the time. Another powerful lobby against Ford was the big banks because he had been a national critic of them.

Even with Secretary Mellon's close ties to ALCOA and President Coolidge, there seemed to be little opposition at the executive level to Ford's proposal. In fact, the Republican administrations of the 1920s preferred privatization of Muscle Shoals. However, Mellon might have lent tacit opposition in the Republican Party in order to tie ALCOA into a deal. The progressive Democrats wanted government control, so in turn, the government could sell energy. Congress passed bills for the government to run Muscle Shoals, only to be vetoed by Republican presidents. Senator Norris of Nebraska led the anti–Ford forces, and the anti–Ford Republican forces often teamed up with the progressives, producing a stalemate. This struggle often pushed the Ford proposal to the sidelines. Eventually, in 1926, Ford gave up in his endless struggle with Congress. ALCOA had also been purchasing land in the nearby rivers in the general area, and they formed Nantahala Power and Light in 1929. ALCOA had been accumulating land, particularly near the Nantahala River, a tributary of the Little Tennessee, and rights to develop hydro power in the Western North Carolina/Eastern Tennessee region in the later 1920s. ALCOA looked to expand in the Muscle Shoals area until the beginning of the Roosevelt administration in 1932, which was particularly anti–ALCOA.

The legacy of Muscle Shoals is sometimes known as the "Alabama Ghost" of the Senate. It was the beginning of decades filled with battles over privatization versus government ownership. Farmers and locals of Muscle Shoals were extremely disappointed. While the government successfully brought cheap power to the South through the government-owned TVA hydroelectric dams, the vision of an aluminum city was lost. In the 1930s, the Roosevelt administration took control of the entire area, and Muscle Shoals became part of the government's Tennessee Valley Power New Deal Project. The other industries of the Ford proposal and the all-aluminum car never came to be. The death of Ford's Muscle Shoals project ended the only serious threat to the American aluminum ingot monopoly.

CHAPTER 10

Paternal Capitalism and Colonialism

IN THE EARLY 1900S, GERMAN-STYLE PATERNALISM received mixed results in American industry. Paternalism was often criticized as welfare capitalism. ALCOA executives could find a wide range of opinions in their East Side Pittsburgh neighborhood on the advantages and disadvantages of paternalism. Carnegie's steel companies, Thomas Mellon's mining, and Henry Clay Frick's mining companies practiced a type of paternalism, which had resulted in violence and bloody labor unrest, similar to problems German companies had experienced. Other East Side neighbors, such as H.J. Heinz and George Westinghouse, had found great success with their style of paternalism. ALCOA would apply a Westinghouse-type plan in more difficult environments, such as the segregated South.

ALCOA had been involved in city building since its early move to New Kensington, Pennsylvania. Its first power projects and mines were in remote areas, requiring the company to bring in workers and supply their needs. ALCOA, by its nature, was one of the world's largest mining companies. These mining areas were remote, and town building was part of the necessary development to obtain workers. In the United States, the mines were in the segregated South, which created a number of challenges. Smelting operations in Canada faced a similar tension between French- and English-speaking citizens. ALCOA was well aware of failures of company towns in Pennsylvania and the famous Pullman City riot in the last part of the 1800s. Andrew Mellon and Richard Mellon wanted to avoid the bad press, and A.V. Davis and Roy Hunt were both extremely community-oriented and were natural paternalists. Many of the communities built by ALCOA would be models for industrial towns. ALCOA proved to have a better record than most industrial companies, but it had its share of problems and critics, too. The labor movement in ALCOA didn't truly gain momentum until the 1930s.

In the first decades of growth, ALCOA, like the nearby steel industry, brought in immigrant laborers from southern Europe such as Italians, Hungarians, and Slovaks. ALCOA proved to be a good employer and community builder, and was able to avoid the labor unrest in the steel industry. The company had dealt with four short strikes prior to 1916, but showed modification, not emulating the steel industry by bringing in scabs. ALCOA had also adopted the practices of Pittsburgher George Westinghouse's company town of Wilmerding. This required a program of Americanization, which educated and prepared immigrant workers to become citizens. The giant steel industry of nearby Pittsburgh had introduced the immigrants to their own resources, often letting immigrant ghettoes develop around the mills. ALCOA's worker-oriented programs at New Kensington kept the aluminum workers out of the bitter national strike of 1919 in heavy industry. ALCOA's approach to workers at New Kensington was clearly superior to that of the surrounding steel and coal industries,[1] and most often exceeded the accepted standards.

ALCOA emulated the best in community building by being similar to the town of Wilmerding. George Westinghouse's town of Wilmerding offered a great contrast to the problems of the area's steel industry towns. Wilmerding was started in 1900, rapidly grew, and became a model adopted by many capitalists. George Westinghouse was a close friend and East Side Pittsburgh neighbor to many of the ALCOA executives. The company town of Wilmerding was of Westinghouse's own design, based on early industrial towns such as Robert Owen's New Lanark in Scotland. In its early years, reporters from all over the world traveled to see a new relationship with industry and community. Wilmerding was hailed by reformers and socialists as the ideal model for industrial towns, but was despised by the bankers and some industrialists of the time, such as J.P. Morgan and Andrew Carnegie, because they viewed it as misguided capitalism. For companies like ALCOA, however, it seemed to be a viable new path.

Wilmerding was in stark contrast to the many surrounding steel towns of Braddock, Homestead, Duquesne, and McKeesport. These steel towns had grown to be unorganized around the mill or mines. There were mud streets where pigs and chickens roamed freely; sewage went in the streets or nearby rivers, and immigrant workers were packed with six to eight individuals in single rooms. Wilmerding, however, was planned properly in every aspect for future growth. Housing was focused on families and six-room cottages, and home ownership was used in place of company-rented apartments. Flowers, birds, insects and hardwood trees prospered, while neighboring steel towns favored only the hearty sumac trees and pigeons. Houses were well built and surrounded by beautiful lawns, parks, and gardens, versus the slag dumps in the neighboring steel towns. Social, educational, and recreational services and opportunities abounded. Citizenship was promoted with the necessary education to support it. A truly symbiotic relationship between company and town existed that was unknown in the Victorian industrial world. While a company town, it was politically independent and open to other citizens and businesses; and New Kensington would match this harmonic corporate model. ALCOA did experiment in paternal capitalism, but ultimately preferred the view of George Westinghouse.

New Kensington, with its location near Pittsburgh, became an independent city as other industries moved in quickly. The Mellon family created a generic town company to support an array of various industries. While it was similar to Wilmerding in that it was politically independent, it had less dependency on ALCOA for maintenance and development. New Kensington was more of a community development project versus a company town; the Mellon family encouraged a balanced array of businesses and industry to support the town, and the Mellon Bank functioned like a land development company by today's standards.

ALCOA's first real challenge in community building was its bauxite operations at Bauxite, Arkansas, where the area had no roads or housing and where racial issues were part of the environment. ALCOA would have to build an entire town from a field and deal with tough racial issues. The work began in 1903 and rental houses were built and utilized for the workers. Within a few years, ALCOA had built three elementary schools and a high school at its own expense. Probably the best feature of the town was its hospital and health service, which was considered to be one of the best in the state in the 1920s. Employees paid a small amount to a health fund and were provided everything for the workers' families. Even in times of unemployment, workers were still given free services at the hospital. Additionally, the company provided assistance with rent and food in these tough times, by far exceeding the northern mining towns in the coal industry, where the loss of employment often meant eviction. Being a Pittsburgh company, ALCOA was aware of the decades of bad publicity attributed to company coal mining towns. The abuses in the coal industry were clearly problematic, and ALCOA wanted to stand apart.

The company also built a theater in town and brought in traveling shows. In 1904, with the cooperation of the Methodist Church, a church was built and maintained by the company. The company built two tennis courts in 1907 and supported a tennis club, revolver club, and boxing club. Most stores, at least initially, were company owned and managed. Overall, the company did work in order to build a successful and comfortable environment. Probably less than half of the 600 workers actually lived in the town; the rest were farmers from the surrounding area who used the recreational opportunities.

In 1915, ALCOA realized that not all was paradise when a strike occurred for better wages, and the company quickly moved to crush it, bringing in outside workers. Historians have noted that while the wages were low, they were to "equal to those in other industries,"[2] and certainly this was true on a regional level. ALCOA was able to settle the strike in a short time by increasing wages. During the war, labor shortages forced the use of immigrants, a majority of Mexican descent, and black laborers. The community became segregated by its own choice, with sections being called Mexico, Africa, and Italy. The company adapted to the ethnic needs of these sections through measures such as publishing a Spanish-language newspaper. Until the 1940s, ALCOA remained the major financial support for the community and proved to be a leader in fighting some of the social norms of the segregated South. In particular, the ALCOA operations eliminated most forms of segregation, but were forced to conform to Southern social norms in the towns. As a northern company, ALCOA considered worker segregation to be counterproductive, but they had a taste of the problem in 1917 at its St. Louis smelting operation, and even earlier, at its first Tennessee mines.

Perhaps ALCOA's earliest challenge in building a fully planned large community was in 1914 with its plant in Alcoa, Tennessee. ALCOA built a smelting plant on a gorge located on the Little Tennessee River, which would supply hydroelectric power. ALCOA would take over the small town of Marysville, in this case. By 1914, the company had completed the initial purchase of 700 acres in North Maryville, Tennessee, and had initiated construction of the smelting plant as well as 150 houses for company employees. These houses were built for families rather than single workers, while dormitories were built for single men and temporary labor. In addition to these new homes, the company financed the purchase of a few existing homes. ALCOA had to build schools to support the influx of families, and as in Bauxite, the schoolteachers were hired by ALCOA. Soon, the company built streets and added utilities. The city was already segregated, and the company adapted to the norm; but the company showed no bias in its distribution of support for its workers. ALCOA's approach to adapt to, rather than change, local norms was a set practice in all its communities. Critics claimed that the company favored blacks because they worked in less favorable conditions and helped resist unionization[3]; and clearly, these factors entered into the management's hiring consideration. The stronger resistance to unions was from the union itself, which refused to allow blacks to join. While ALCOA adapted to segregated living, all benefits were applied on an equal basis. ALCOA leased lots to three new churches, and one included the all-black St. Paul African Church. Black schools were built with the same high-quality standards that were applied to ALCOA's white schools. In 1920, the town's population consisted of 1,708 whites, 1,482 blacks, and 150 Mexicans.

The one type of company bias that did exist was the belief that blacks were able to withstand more heat; and because of that, they were assigned to the smelting plant. The belief may have been wishful thinking on the part of company managers, since the smelting plant was the dirtiest and hottest environment; and many Southern white workers refused to work there. Mining proved to be favored by Mexican labor, and generally only the Mexicans would deep mine for the wages offered. Still, the pay was equivalent for black, Mexican, and white workers, and the pay was much higher than other work available to them in the region. In the 1920s and 1930s, the company was competing with better-paying jobs for blacks in the Detroit auto industry, and therefore, benefits and pay were increased. The town did, in fact, lose a significant number of black employees to Detroit, but ALCOA wages reversed the trend by 1929. These workers represented a major investment by the company, and operations made every effort for equality of races within the plant. Outside of the plant operation, however, the company did conform to pressure from the Democratic Party machine to restrict opportunities for blacks. An example was the hiring of black schoolteachers at lower pay.

A recent historian noted: "Even within the confinements of segregation, however, there were some differences in ALCOA. Company officials believed that African American workers also deserved quality community life, even in the separate [black] Hall and Oilfield communities. Arthur Vining Davis, the company's powerful board chairman, wanted the best African American school system in the South for the company's laborers."[4] ALCOA's famous citizen and U.S. Senator Lamar Alexander also stated that even with segregation, the company treated the blacks fairly. Blacks, like whites, were retained in slow economic times with menial jobs and help with the rent. Another motivation was to keep blacks from migrating to higher-paying jobs available for them at Ford Motor Company in Detroit. Black workers had proven to be an asset in the hot and dirty smelting department. Lamar Alexander further noted in a foreword to the history of the town: "During this last century, ALCOA has become much more than a company town. But a legacy of the aluminum makers from Pittsburgh is one we should honor."[5] Also, this is a credit to A.V. Davis, who improved on the paternal practices of American capitalism. But, of course, ALCOA was not without its race problems in the South.

ALCOA found even bigger problems with race relations in the North. The East St. Louis plant was a hotbed of race tensions in 1917. As a border state, East St. Louis had a history of race problems, but had remained predominantly white for decades. Prosperity had created a need for workers at the plant, and East St. Louis had become a staging area for the great migration of blacks. The migration was a result of the collapse of the cotton industry with the infestation of the boll weevil. Thousands of African Americans traveled north to fill jobs in the booming Northern industries. The city of East St. Louis, Illinois, would be the scene of one of the bloodiest race riots in the 20th century. Racial tensions began to increase in February 1917 when 470 African American workers were hired to replace white workers who had gone on strike against ALCOA's Aluminum Ore Company. The strike lasted only a month, but it increased racial tensions amongst the striking white union members as ALCOA continued to hire black employees.

Estimates of between 10,000 and 12,000 African Americans left the south for East St. Louis in 1916 and 1917 as part of this industrial migration. The mass migration was due in part to the boll weevil and the destruction of cotton plantations. The migration brought in blacks who would work at lower wages. Many white citizens of East St. Louis were disturbed by this movement and by the increase in employment of black people at the ALCOA plant. By summer, there were nightly riots and crimes against blacks. The National Guard was called in but lacked the will to do much. The riot hit its peak on July 2, 1917, and violence took hold of the city. ALCOA then found itself in the middle of a race riot as a major black employer. ALCOA's plant gave asylum to frightened blacks and some employees were armed with rifles. A combination of police and military personnel were finally able to bring an end to the violence, but not before the death toll rose to somewhere between 50 and 200.[6] ALCOA maintained its policy of hiring African Americans due to the shortage of workers, but also increased employment overall. ALCOA had learned a great deal in terms of community relations from the riot, and the information would be applied to its Southern bauxite operations.

ALCOA would find relations between French- and English-speaking workers in Canada just as challenging. Arvida was founded as an industrial city by ALCOA in 1927, when the first aluminum smelter was constructed in Canada's Quebec. Its name is derived from the name of its founder, Arthur Vining Davis, president of the ALCOA aluminum company. Arvida was located 150 miles north of Quebec City and south of the Saguenay River between Chicoutimi and Jonquière. The area was sparsely populated, but there was a rail station that needed to be improved for massive shipments of raw materials, so ALCOA had to invest heavily to expand rail service. Arvida was located in the heart of French-speaking Quebec and was known as "the City Built in 135 Days," described by the *New York Times* as a "model town for working families." The town was planned from the first day and was developed as a company town to have an initial population of about 14,000 inhabitants. The town was built up quickly, but was very well planned. It would have four Catholic churches and parishes, along with churches of many other denominations, and schools. Arvida was a planned ALCOA community more like New Kensington compared to ALCOA's southern mining operations.

The Arvida location was remote, but ALCOA planned to send its best young managers and their families to the area. The remoteness would hit home when the temperature was –42 degrees F on Christmas Day. During the winter, most of the main roads to major cities, such as Quebec City, were closed, and only a few managers tried to operate cars in such extreme cold conditions. Horse-drawn wagons and dog sleds were the most common means of transportation, which was quite a drastic change for many Pittsburgh-born managers. This remoteness necessitated a town with good schools, parks, roads, recreation centers, etc.; and so the prominent New York architects, Hjalmar Skouger and Gamble Rogers, were hired to design the town. While originally the town was to have 3,000 citizens, the plan called for a future city of 30,000. The original houses were offered for sale or rent. Later, the town offered all houses for sale, ending its role of landlord. ALCOA had always preferred the Westinghouse model of selling homes instead of the failed rental model of Pullman City in Chicago and Pennsylvania coal mining towns. The company built a hospital along with extensive recreation facilities. The town was to be self-governed with no political involvement by the company. Like Westinghouse's Wilmerding, the independent city still used the company resources extensively. The American managers and their families were true pioneers in building Arvida. The French-speaking managers and workers proved just as hearty, often coming to work via dog sleds. However, ALCOA eventually found the bigger problem caused by mixing French-speaking Canadians with Americans.

The bilingual nature of the area and factory required a set plan of corporate diversity. The French-speaking Canadians demanded the use of French. ALCOA, and later Alcan, became leaders in the adoption of French as the working language of industry, and newspapers, signs, and work instructions were printed in both languages. French-speaking middle managers and foremen were hired for supervisor positions. Efforts were even made to integrate French- and English-speaking workers while respecting their own heritage. While the town was distinctly French Canadian, concessions had to be made for the transplanted American managers by adding French-speaking assistants for those managers. A country club was built to address the passions of many American managers. An amusing story is told of a visit to Arvida by board member Richard Mellon. The tour guide recalled: "As we moved along,

Richard Mellon said to Mr. Davis, 'Arthur, what is the name of this street?' Harold Wake said 'Oersted Street.' Immediately, Mr. Davis said, 'It's Mellon Street,' and turning to Wake said under his breath, 'Get that name changed right away.'"[7] Another street was called Oersted, and today, both streets still exist.

At the core of ALCOA's success with town building was a positive approach to employee work relations and Davis's personal belief in community building. From its beginning, ALCOA showed an enlightened approach to employees and community. Hall and Hunt clearly adopted the practices of their neighbors on the East Side of Pittsburgh, George Westinghouse and H.J. Heinz. There was a deep belief that productivity was more employee-oriented than technologically-oriented. In 1902, the Pittsburgh Reduction Company board created a "Permanent Employment Fund."[8] The plan was visionary and generous to a fault as the fund functioned like a pension and disability plan. The company credited every hourly employee with 2½ percent of his monthly earnings to the fund. The company added a six percent interest to the fund semiannually, and the employee was required to add nothing. It was initiated at the Niagara Falls plant, followed by the plants at New Kensington, Massena, and East St. Louis. The hope was to create a stable long-term workforce because the immigrant workforce available to industries of the time was quite transient. Young, single workers often worked a few years and returned to the home country, while workers in their 20s often would leave after 8 or more years of saving to start a small business or move into skilled trades. The ALCOA plan operated with the goal of keeping those younger and family-oriented workers.

The Permanent Employment Fund, although a visionary idea, had its flaws. One of those flaws was that the plan was too generous. If an employee gave 30 days' notice and left, the employee received ownership and interest on the dollars credited to him. This defeated the purpose and gave the employee little incentive to stay. In fact, the nest egg built up actually worked as incentive for some employees to leave in order to access the fund. The initial plan was adjusted, but this measure proved ineffective; and when employees at Niagara wanted a wage increase, the company discontinued the plan in lieu of a raise.

Later, ALCOA's strategy was to offer security for the worker against his biggest fear—loss of work due to the economy or due to sickness. Also in 1902, along with the Permanent Employment Fund, Pittsburgh Reduction created a "Sick Fund." This was a contributory type fund where the employee gave 25 cents a week that was then matched by the company. This plan worked well, and was popular with the workers; the sick fund would eventually evolve into company insurance programs. The recession and auto depression of 1920–1921 created new fears of loss of income. The employment at ALCOA went from 20,952 to 10,007 during the 1921 auto recession. The company in response set up a voluntary Aid and Benefit System. Each dollar the employee donated, the company matched. This was novel non-government unemployment insurance for the times. The plan was popular and offered relief during the Great Depression of the 1930s.

CHAPTER 11

The 1930s Unionization and the End of Paternalism

THE 1930S BROUGHT MAJOR CHANGE TO LABOR RELATIONS in the United States. The business-friendly Republican administration of many decades was gone, and it was replaced by the progressive and socialistic administration of Franklin D. Roosevelt. The National Labor Relations Act, or Wagner Act, named after Senator Robert Wagner of New York, was the cornerstone of President Roosevelt's New Deal social legislation. This Act changed the very landscape of labor in the United States in hopes of avoiding widespread social unrest during the Great Depression. It would also promote a wave of unionization in American heavy industry. Companies turned many of their responsibilities over to government programs and union initiatives. The 1930s depression would also leave a lasting mental scar on the American worker.

The stock market crash of 1929 resulted in over $2.2 billion being called on stock loans, and those losses put many in the nation in desperate financial positions. The crash was not directly followed by a depression, but it did signal the end of a huge expansion and bubble in the economy. By the end of 1930 and into 1931, bank failures, panics, and runs started with the now-famous lines of panicked depositors. The banking problems spread in 1931 with a doubling of bank failures and mortgage bankruptcies. By 1933, 11,000 of America's 25,000 banks had failed, and companies stopped hiring and began reducing their workforces.

The money crisis soon spread to other parts of the economy. For example, home building dropped by 80 percent and factories were forced to cut production. Factories changed to a schedule of two to three days a week, effectively cutting wages 50 percent. Social problems started to mount as unemployment hit 17 percent in 1931 and would peak at over 25 percent of all workers and 37 percent for non–farm workers. Even these estimated unemployment rates understated the panic since the majority of most Americans were working short weeks. In 1931, ALCOA was forced to reduce wages and salaries; ultimately, the effort fell short and the company laid off 23 percent of its workforce. In rural communities, where the community was the workforce, ALCOA reduced hours to avoid layoffs.

The Great Depression changed the American psyche and the economic thinking of the American worker. The loss of work and massive unemployment rates showed that paternal capitalism could not handle the size of the economic distress. However, despite the depression, Americans remained patriotic and continued to show support for the government, whereas the poor economic conditions contributed to growing social unrest throughout Europe. The family unit was key to America's weathering of the depression. Smaller towns, like Alcoa, Tennessee, showed how a strong community could survive such a downturn, and ALCOA stepped up its support to communities. The depression would also initiate a decade of social and economic legislation to build a safety net. In August 14, 1935, President Roosevelt signed the Social Security Act, which was a social insurance designed to pay retired workers age 65 or older a continuing income after retirement. It was not originally planned to be a full retirement fund, but rather to maintain the income of the retirees above poverty levels. In 1937, the payroll taxes were one percent for the employer and one percent for the employee. Payroll deductions were mandated under the Federal Insurance Contribution Act (FICA). Workers turned to the government and unions for more support, while pressuring companies for better wages.

Roosevelt hoped that higher wages obtained by unionization would help improve the economy. He used executive power and threats as well as legislation to push for unionization, and the National Recovery Act started the evolvement in 1933. The National Labor Relations Act of 1935 (Wagner Act), which strengthened Section 7 of the National Recovery Act, would usher in a wave of unionization in the steel, automotive, rubber, mining, electrical, and aluminum industries. At the start of the 1930s, less than 10 percent of America's labor force was unionized; but after the passage of the National Labor Relations Act, the decade ended with 35 percent of the labor force unionized. Once the National Labor Relations Act of 1935 was on the books, it offered a means to push unionization in America's biggest industries. The Act was, to some degree, an overreaction, but balance would be achieved after years of debate and amending. The ultimate goal of the Act was to improve the economy by increasing wages, but that was never fully achieved because wage increases went hand in hand with a reduction of benefits and community aid. Unions started to take over the paternalistic role of companies.

The heart of the Act is Section 7: "Employees shall have the right to form, join, or assist labor organizations, to bargain collectively through representatives of their own choosing, and to engage in other concerted activities for the purpose of collective bargaining or other mutual aid and protection." The key was a secret ballot election of the union under the auspices of the National Labor Relations Board (NLRB). Now, a procedure backed by the government was in place.

Section 7 also permitted the establishment of standards regarding maximum hours of labor, minimum rates of pay, and working conditions in the industries covered by the codes. Additionally, Section 7 authorized the president to impose such standards on companies when voluntary agreement could not be reached. ALCOA and other companies found their wage structures and benefits under the review of the National Recovery Administration Policy Board. The industrial environment was now ready for the union movement to succeed against paternal capitalism.

The American Federation of Labor (AFL) had evolved as a crafts union for skilled white workers. It was the creation of Samuel Gompers in the 1800s. Its crafts orientation favored groups like machinists, foundry workers, and other similar trades. The efforts to unionize prior to 1930 in the steel industry had failed because of its resistance to include unskilled workers. The early American Federation of Labor unions (AFL) clearly discriminated against blacks and people of color more so than the companies. The National Recovery Act clearly favored the Committee of Industrial Organization (CIO) unions over the AFL because the CIO was all-inclusive.

The AFL, for the most part, had a record of failure in organizing heavy industry. It had made inroads for recognition, but with mixed results. The AFL was a very conservative organization, resisting strikes as much as possible. However, its bias against blacks and unskilled workers often resulted in a lack of solidarity during strikes. The AFL failed to adapt to the change in industry from skilled to unskilled labor driven by the technology. Heavy industries such as steel, automotive, and aluminum, had beaten back the unions for decades. The role of the AFL in the National Strike of 1919 pointed to AFL's failure to manage strikes in heavy industry. The steel companies broke strikes by using 30,000 to 40,000 black and Mexican-American laborers. The lack of solidarity between white and nonwhite workers prevented successful strikes and resulted in major defeats, such as the National Steel Strike of 1919. The AFL's own bias against nonwhite workers was part of the 1919 defeat and, from then on, the AFL was hesitant to call for strikes. Of course, the lack of government support for unions in the 1920s also contributed to the decline of the AFL. Then, the 1930s and the high unemployment rates created anxiety in workers. They could no longer count on companies to help in such a deep depression, and workers needed to look elsewhere for long-term security.

The CIO had formed to take on large U.S. companies in 1935 and take advantage of the Wagner Act. The CIO was in its embryonic stages of bringing together eight international unions under the guidance of committee leader John Lewis. Lewis was from the United Mine Workers, the most successful industrial union. The union movement would take a new direction that would focus on inclusiveness and solidarity and, overall, would unite the whole workforce at industrial plants.

It was clear that unions would have to move from a crafts model to an industrial model. Skilled and unskilled laborers and black and white laborers could unite under one union. Solidarity was needed to win strikes and negotiate wage increases, which required opening membership to all workers, even African Americans. Roosevelt tried to threaten big companies such as ALCOA with antitrust actions. Unions had rallied under Roosevelt in 1933, but strikes proved unsuccessful except against the mining industry. After much blood and violence, the miners won recognition for their union by the mining companies. As laborers came together in CIO solidarity, the core model to expand the CIO into other major industries was set. In 1936, the rubber industry was unionized by the CIO.

Within a year of the rubber strike, the Steel Workers Organizing Committee (SWOC) of the CIO evolved. On January 1, 1937, United States Steel accepted a one-year contract with the SWOC that included union recognition, the 40-hour work week, time and a half for overtime, holiday pay, a week of vacation, and a $5 per day wage. Other steel companies thought United States Steel had sold out and, as a result, a national strike against those companies raged in 1937. The strike reached a peak at Republic Steel's South Chicago Plant, where 10 workers were killed in one day (7 shot in the back) and more than 50 workers were wounded by police. This bloody strike ended with the acceptance of the United Steel Workers union throughout the steel industry. Thus, in 1937, the first industrial CIO union was fully developed. The Act's first major success for raising workers' wages through unionization came in 1937 with the victory of the Congress of Industrial Organizations (CIO) union over General Motors. ALCOA was slow to be unionized because it was manned by AFL franchises, had a good labor track record, and overall provided good wages and benefits. The success of the CIO challenged the American Federation of Labor, causing a split in the union movement that would affect ALCOA.

ALCOA's issue with unionization was not with wages or benefits, since aluminum manufacturing was far less labor intensive than industries like steel. The issue for the company, and to a large extent for the workers, was control of the factory floor. The paternal system had weaknesses, such as favoritism of certain workers, and a problem with a foreman could significantly impact an individual worker. The company still had control of exactly how and how long an employee worked. In hard times, the company was in complete control of who worked and who didn't. The company believed it needed flexibility in how work was managed, while most capitalists of the period feared this loss of control.

The establishment of large industrial unions and the implementation of government social programs, such as Social Security and unemployment, ended the application of many paternalistic practices. Additionally, programs and plans would end, like ALCOA's Sick Fund and wage protection funds. Wages and benefits would be settled by collective bargaining. The union also ended the practices of control of the worker by the company, and the ability to fire and assign jobs came under the scrutiny of the union. ALCOA tried to avoid any violence and took a watch-and-see approach as the steel industry, auto industry, and rubber industry were the forefront of the union movement in the mid–1930s. ALCOA had experienced strikes, but avoided most of the violent strikes that occurred in other industries.

Pittsburgh Reduction Company's first taste of employee relation and labor issues came in March of 1900. A group of 37 employees from the wire mill walked out and demanded an increase of 50 cents from their hourly rate of $1.25 an hour. The men gained another 100 employees and demanded to be recognized as a union. The walkout lasted a few days with the men receiving a 15-cent increase and recognition as the Aluminum Workers Union, AFL Local 8261, although it would be a short-lived union. In 1907, Pittsburgh Reduction broke the union and the workers rescinded their membership. There were few walkouts prior to 1919, but none proved to be successful. Most of this was their proactive approach to labor issues.

Pittsburgh Reduction had learned from the steel strikes of Pittsburgh to avoid violence in 1919, when national strikes hit the mining and steel industries. First, the company was proactive with its benefit programs, which were far superior to those provided for the steelworkers. The company's average yearly wage was $1,169, compared to the all-manufacturing wage of $1,140 in 1920. Secondly, management at New Kensington, Niagara Falls, and East St. Louis was open to listening to employee demands. Thirdly, the company preferred to apply pressure on individuals by selective firings and work assignments. The Slav and Polish immigrants were the most problematic in terms of labor unrest prior to 1920, but ALCOA embraced the immigrants. ALCOA used the Westinghouse model in the city of Wilmerding and offered education and citizenship courses. This Americanization program worked well against the socialistic International Workers of the World (IWW). Foremen were actively engaged in crushing the seeds of unionization. Good workers were encouraged and got more favorable treatment. Still, organizers were harassed.[1] The relationship of community and company was a tight one for ALCOA. New Kensington gained a reputation of an anti-union town, as other local industries also resisted unionization.

The workers at the ALCOA mines fared much worse than the aluminum manufacturing plants on a straight comparison, but stood out for better wages and treatment from competing Southern industries. Much of this was due to the inability of unions to include black and Mexican workers, because of the racial bias of Southern-based companies. The work was also backbreaking and dangerous. The mines, however, were at remote locations dependent on the company. The company towns with imported cheap Mexican labor and regional black labor were looking for better work than the menial jobs offered by other Southern industries. Both groups of workers were unacceptable to unionizers at the time, and the company afforded better treatment for Mexicans and blacks than might be expected in a union environment. Similar to ALCOA's northern strategy, the strategy at the mines was to be better than the regional average in wages and benefits. ALCOA's paternal approach had won the workers over at most plants.

Pre–World War I, ALCOA faced a number of short strikes at New Kensington over working conditions. The most serious demand was the elimination of the ten-hour day and implementation of an eight-hour day. The eight-hour day had been one of the major demands of the Knights of Labor in the 1880s and had been at the heart of the Haymarket Riot in the 1880s. Still, the ten-hour day remained the standard of the 1800s. At the beginning of the 20th century, the American Federation of Labor (AFL) joined the movement during a time when heavy industry remained steadfast in the ten-hour day. An unaffiliated union movement took up the eight-hour fight at New Kensington in 1916. The walkout of machinists would last only two weeks as ALCOA management waited it out. It would be in the 1930s when ALCOA would face its true union challenge.

The full impact of the 1929 stock market crash hit America's factories in 1931 and ALCOA was hit hard. ALCOA initially reacted by reducing wages and salaries, and employees worked a reduced work week for months in an effort to avoid heavy layoffs. However, by the end of 1931, ALCOA was forced to reduce the workforce by 23 percent, although the cut was smaller than most of the cuts within the heavy industry, and less than the 50 percent reduction that occurred during the earlier 1920–1921 auto recession. The poor economy set up a favorable environment for unionization, but the union movement was split between the American Federation of Labor (AFL), the Committee for Industrial Organization (CIO), and various socialist activists. The New Kensington plant was in the heart of the union movement. Within a 100-mile radius, there were the steelworkers to the south, rubber workers to the west, and miners to the east. New Kensington was slow to join in with large numbers due to a degree of contentment with the paternalism of the company, but eventually the depression spread fear and anxiety. The long-term reduction of pay put pressure on the workers and their families that had not existed before. While the workers were mainly focused on wages and security, the activists were interested in union recognition. All in all, at the time, ALCOA management was able to hold firm against unionization throughout the area.

The National Recovery Act of 1933 challenged ALCOA's strategy of maintaining the status quo, as ALCOA now had to deal with the government. Having full control of the aluminum industry, ALCOA was already under review by the Treasury Department and was not in a position to resist the government on another front. Section 7 of the Recovery Act assured the workers that ALCOA would have to address their concerns. Roy Hunt, president of ALCOA, hoped to avoid confrontation with the government and union organizers by proposing his own plan to meet the concerns of National Recovery Policy Board. Roy Hunt was one of ALCOA's most progressive managers, but like most at the time, he opposed unions. Both Roy Hunt and ALCOA had a strong record of good worker relationships.

The Recovery Act plan was well designed to offer slightly higher wages while remaining below the expected demands of the workers. The plan centered on wages which were acceptable, with minor changes to the Policy Board. ALCOA tried to model a company-type union after the union at Goodyear Rubber. The Recovery Act required the approval of workers by secret ballot, and in 1933, the workers at New Kensington voted the plan down. At the same time, the workers formed a local chapter of the AFL, known as Local 18356. AFL would also organize unions at other ALCOA plants, such as Massena and East St. Louis. Under the Recovery Act, ALCOA had to accept the union, but they tried to minimize its impact and used its community clout to resist active union participation. It was even reported that New Kensington merchants refused credit to union supporters.[2]

The first test of the new union came in 1933 with a walkout of 5,000 workers at New Kensington. One problem with the AFL was that the factionalized organization experienced power struggles at the national and regional levels. AFL's first priority was to establish a strong core by collecting dues. Because the national organization was in no position to support local strikes, ALCOA took advantage of that weakness, and the ten-day strike at New Kensington broke down. The union achieved an across-the-board pay raise, but failed to have the union formally recognized, and did not take union dues out of paychecks. Roy Hunt proved to be a tough negotiator as he became an expert of the Recovery Act who exploited every weak point without exploiting the workers. Hunt was successful at getting support from the National Recovery Act Policy Board for a basic national wage of 40 cents an hour for males and 35 cents an hour for females. The AFL complained that those wages undercut their own efforts, and when the national AFL refused to fully support the Local 18356, it led to a slow internal death of the AFL Local union at New Kensington.

Local 18356 was led by fiery union leaders, which included a woman named Mary Peli. ALCOA had been one of the first in heavy industry to hire women. Peli told a much different story about the working conditions at New Kensington than most. Hundreds of women worked at ALCOA and represented a majority in many departments. Mary was particularly upset by the hiring practices of women. Her claims included: "Each day thousands of women would appear at the plant and were forced to endure the company's version of the shape up. The women were lined up and inspected military fashion. Those who were hired were admitted to the shops and were subjected to harassment and intimidation."[3] Mary Peli's claims have some justification, but the first strike was strictly over low wages and preferential hiring and firings.

In August of 1934, the AFL tried to stage a national aluminum workers' strike. The effort focused on union recognition, layoffs by seniority, and universal industry wages. This time, the strike lasted five weeks, but the poor economy and planned stockpiling by the company put management in control. The will and solidarity of the union workers were weak. The company refused to go to arbitration by the newly formed National Labor Relations Board, but did allow mediation with the Federal Labor Department. Some minor gains were made, but overall, the workers were dissatisfied with the AFL. New Kensington was particularly rebellious, with an active socialist element that clashed with the very conservative approach of the AFL. The internal struggle at New Kensington became a major problem for the next few years.

The AFL and Local 18356 were clearly outmatched by the company, and ALCOA was winning without using violence or lockouts. Some of the company's success was due to the paternal approach and community building. The company did apply some tried and true strong-arm methods such as harassment and community pressure, and it used moderate wage increases to build anti-union sentiment. Additionally, the company offered the union president a managerial position, a common strategy in heavy industry, and he accepted. The union was divided and most members had serious problems with AFL leadership. The AFL refused to give Local 18356 an international charter, and the local was viewed by the other AFL unions in ALCOA as radical socialists. The

AFL Local 18356 had a membership of 3,300 at the time of the 1934 strike, but it was down to only 17 members in 1935. The local seemed to be at odds with everyone and was in disarray with officers fighting one another. There was also internal corruption that occurred as some dues went missing. Many believed that Local 18356 and the AFL were the real obstacles to unionization, not ALCOA.

In 1935, the National Recovery Act was struck down by the Supreme Court. New Deal democrats quickly passed the Wagner Act, which strengthened the union movement; and in 1937, the Committee for Industrial Organization (CIO) was born. The CIO was clearly more radical than the AFL and was very aggressive with wage demands. In 1937, the CIO had major victories in the rubber, auto, and steel industries as significant wages and benefits were achieved without bloody strikes. However, the formation of the CIO created a war with the AFL over union membership, and ALCOA played the division perfectly. It raised wages again and formally accepted AFL as the bargaining agent, playing to its weaknesses; therefore, most ALCOA plants were represented by the AFL. At New Kensington, Local 18356 realized that the company could not be moved without a more aggressive approach.

To the east of New Kensington, the rubber workers in Ohio dropped the AFL and moved to the new CIO. The AFL, like previous unions, had a record of failed strikes in the rubber industry. The AFL tried to hold on, but in 1935, the formation of the United Rubber Workers came under the guidance of the CIO. In 1936, the CIO won huge strikes against the rubber industry. Then, the automotive and steel industries shadowed by turning to the CIO.

Local 18356 quickly followed the lead of the auto and steel industries in formally aligning with the CIO while ALCOA unions remained firmly with the AFL. New Kensington became a major battlefield for the AFL and CIO. The battle would soon become an internal one, and behind the scenes, ALCOA and Roy Hunt were pushing for the AFL. The CIO felt its best chance to overtake the AFL in the aluminum industry was to flip the radical Local 18356. The alignment with the CIO did bring back thousands of members who had been lost in the two years previous. A vote in April of 1937 was a major victory for the CIO as AFL Local 18356 was dissolved and replaced by the International Union, Aluminum Workers of America, CIO Local 2. While the AFL would sue in court, the National Labor Relations Board certified Local 2 of the CIO as the bargaining agent. Still, other ALCOA plants continued to stand strong with the AFL.

The next CIO-AFL battlefield would be at Alcoa, Tennessee. The problem in Tennessee was similar to that in New Kensington in that there was a struggle in the local union. ALCOA was having issues dealing with the Tennessee workers at the ALCOA mines because the workers had objected to the company plan filed with the National Recovery Administration Policy Board. The plan discriminated against Southern workers by paying them 35 cents an hour versus 45 cents an hour for northern workers. The wages were high on a regional basis. The AFL had organized the workers in 1933, forming Local 19104, probably with the blessing of ALCOA management. The AFL had a policy of not striking over north-south differences in pay. The AFL also treated the Tennessee plant as a special case, allowing unskilled and black workers to join, which was a major plus. Although blacks were not actively recruited for membership, some did join, and of course, the practice including more individuals strengthened the solidarity of the union when scabs could be used to break strikes. The company seemed also to treat it as a special case by giving the AFL recognition and the right to collectively bargain.

Alcoa, Tennessee, was one of ALCOA's best company towns. Living conditions were superior to those found in other towns in the region. Its school for both black and white students was considered one of the best in the state and the wages were above the state average and far superior to farm labor. The setting would hardly seem to be the location of the company's only fatal encounter with striking workers. Yet it was in Alcoa, Tennessee, and not with the more radical unions at New Kensington and East St. Louis, where the fatal encounter occurred in 1937. In July of that year, the AFL reluctantly supported Local 19104 in a strike over wages. When the company responded with scab labor, the union turned to attacks on company property. Police were recruited from the surrounding counties and the confrontation led to the killing of one policeman and a worker. Eventually, the National Guard was called in and the strike was broken, but ALCOA's public image was seriously damaged. As a result, the company was exposed to pressure from the Roosevelt administration.

The national president of the AFL came to ALCOA to ask for the resignation of the local president. The local workers felt betrayed by the AFL, and the company sponsored its own type of union called the Aluminum Employees Association, which grew quickly to outnumber the AFL Local. The company union seemed to gain support from not being connected with outsiders. The wages were increased to 45 cents an hour as a short-lived economic boom occurred in mining. In 1937, with no violence or strikes, the CIO won the right to bargain for United States Steel. This was a huge victory for the CIO and put pressure on the aluminum workers to join. The CIO attempted a number of times to organize, and after four NLRB elections, the CIO became the bargaining agent in 1941. ALCOA then signed a national bargaining arrangement with the CIO, which represented over 50 percent of ALCOA's operations.

The CIO did take some steps to improve the treatment of blacks. The union hall was integrated, although workplace restrooms, water fountains, and cafeterias remained segregated. Some historians argue that black union benefits were not equal[4] because the union, like the company, was restricted by local norms, but both did advance the condition of blacks in the workplace.

On a national level, ALCOA plants were split between the AFL and CIO. By the end of the 1930s, only Massena, New York, and East St. Louis plants were represented by the AFL. The CIO challenges continued even at these two plants, but with limited success. The struggle between the two unions continued through the war years. The power on a national level grew for the CIO, and in 1944 the 45,000-member Aluminum Workers of America (CIO) merged into the United States Steelworkers. ALCOA and the industry in general had avoided the labor problems of other heavy industries.

CHAPTER 12

The Great Antitrust Case

THE 1930S PROVED TO BE MORE CHALLENGING for ALCOA and not only because of labor problems. The political environment had fundamentally changed from the pro-business 1920s and the decade-long rule of Secretary of the Treasury Andrew Mellon. Mellon and his association with ALCOA became the point of attack for populists and New Dealers in Congress. Mellon was portrayed as second only to Herbert Hoover as the scapegoat for the depression, while the press supported viewing Andrew Mellon as a villain. Franklin Roosevelt ran on these views, and after winning the election, he continued to look at ways to punish Andrew Mellon. Mellon and ALCOA both became targets of this populist retribution, and the Roosevelt administration used the Internal Revenue Service to bring charges against Andrew Mellon personally. The IRS claimed tax fraud and the trial went on for a total of four months. In the end, Mellon was completely absolved of any wrongdoing. However, one of Mellon's biographers would call it "the first [battle] in an all-out war with the Roosevelt administration."[1]

For the Roosevelt administration, the battle against Mellon was not over. It continued to be a bitter smear campaign in the press against Mellon as Roosevelt still had the press and public behind him to further attack Mellon and "his" ALCOA. The Roosevelt Administration would focus on what was considered America's biggest monopoly, ALCOA. First, the new political environment would bring a series of private antitrust suits. George Haskell, having lost previous antitrust cases, thought the time was right again in 1933 with the new administration. Haskell Manufacturing was an independent fabricator of aluminum and often bought cheaper foreign ingot in order to be competitive with ALCOA's vertically integrated operations. Again, Haskell alleged that ALCOA had willfully undercut prices on his Brush Machine Company sheet and forgings. Furthermore, he claimed that ALCOA had conspired with foreign producers, and he was asking for $3 million in damages (five times ALCOA's net income in 1933). Once again, ALCOA was cleared of all charges by a Connecticut jury. In 1935, a federal appeals court overturned the decision, but in return, ALCOA appealed and won again. Haskell threatened yet another suit as Roosevelt was aggressively packing courts to support New Deal legislation. With this particular suit, ALCOA settled out of court and placed Haskell on retainer. The Roosevelt Justice Department was far from placated, though, because the ultimate desire was to break up the aluminum monopoly. Previous antitrust cases were reviewed and the Justice Department would bring new charges in 1937.

For years, ALCOA had been considered a necessary monopoly to bring this critical metal to market. Its role in World War I had been part of the nation's success, and because of that role, ALCOA was allowed to operate with less scrutiny by the Justice Department. ALCOA was far more monopolistic that AT&T, Standard Oil, and United States Steel, but for decades had enjoyed this monopolistic position without any limiting antitrust decrees. No other industry required such an international combination of resources, which also made it a strategic company to the nation's defense. Aluminum, by World War II, was more strategic than rubber or steel. The management of this industry required a unified and vertically integrated approach to be efficient. In Europe, that efficiency was gained through cartels, but in America it was a national monopoly. ALCOA was a monopoly, but even more, it was the core of the nation's aluminum industry and a strategic defense advantage. Initially, it was ALCOA that had to develop products and processing industries to deliver these products to market. On many occasions, ALCOA cut the secondary manufacturing operations free because of the overextension of its management, and this helped pacify the Republican administrations of the first quarter of the 20th century. In the 1930s, ALCOA avoided making any moves that the Justice Department may view as wrong and use it against them.

From 1935 to 1940, ALCOA maintained a price reduction. The price of ingot was 15 to 20 cents a pound, while ALCOA's production cost was about 10 cents a pound.[2] They maintained a decent profit without raising the price to that of foreign producers, and the lower ingot price would become a major barrier for entry by new companies. The Roosevelt administration believed it was done because of the antitrust suit. I.W. Wilson, vice-president of ALCOA operations, had also increased capacity to 700,000,000 pounds. As the nation moved toward war, Wilson assured the government that ALCOA could meet warfare demands without the need for their interference. It now appears that ALCOA overestimated its ability to supply the military as a way to keep others from government-supported industry growth. The military was the only part of government that preferred the monopoly position of ALCOA.

Aluminum innovator and famous engineer Buckminster Fuller took his view of ALCOA a step further. Fuller stated: "Suddenly, we had a new form of capitalism, which required both large scale financing and integration of metals, mines, and mine-owners, metal refining and shaping into wholesale forms, all to be established around the world by the world masters of the great line of supply. The world line of metals-and-alloy supply was essential in producing all the extraordinary productive new machinery and that machinery's delivery system, as was the generation and delivery of the unprecedentedly vast amounts of inanimate energy as electricity."[3] The sheer quantity of energy that was required caused it to be called "packaged electricity" or "frozen energy."

ALCOA and Andrew Mellon continued to have a major challenge in terms of their image during the depression. Capitalism was an easy target in bad times, and the press and populists resorted to class warfare. When New York papers statistically claimed in 1930 that the nation's 60,000 wealthiest families had as much money as America's 25 million poorest families, Franklin D. Roosevelt (then governor of New York) called Andrew Mellon "the mastermind among malefactors of great wealth."[4] The main argument was that families like Mellon achieved this wealth at the expense of the lower class. The argument played extremely well outside America's manufacturing cities and became the strength of New Deal politicians. Right or wrong, such an argument created an image problem for ALCOA. The negative image was somewhat modified, however, as the country moved toward World War II, because aluminum was a major American advantage.

The military was dependent upon ALCOA to develop the necessary alloys for demanding war applications. During World War II, Europe used centralized government research to create new alloys and weapons. America, with its capitalistic system, preferred private industry to handle research. President Woodrow Wilson in World War I did consider nationalizing big industry, but its failed effort in nationalized shipbuilding showed the government to be an inefficient manager of big business. The late 1930s required a balancing act by the Roosevelt administration to limit the ALCOA monopoly, while at the same time building the defense industry.

For decades, it was clear that ALCOA violated the Sherman Antitrust Act, requiring even business-friendly administrations to investigate ALCOA in the 1920s. Ultimately, the antitrust cases cleared ALCOA of all wrongdoing. Many saw the hand of Andrew Mellon as Secretary of the Treasury in the perceived success of ALCOA in the investigation outcomes. Andrew Mellon had given his stock to his brother Richard when he became Secretary of State, but the Mellons were a tight financial entity as a family. Still, the monopoly position of ALCOA had a long history.

In 1911, the government had used the Sherman Antitrust Act to break up Standard Oil. The importance of the decision *Standard Oil Company of New Jersey v. United States* was its use of the "rule of reason." The rule of reason stated that large size and market control were not necessarily bad or violations of the Sherman Antitrust law, but it was the use of abusive tactics to attain and preserve such a monopoly that was illegal. The concept of rule of reason appealed to the business community as a fair ruling and something all could live with. The rule of reason had protected ALCOA because investigators had to prove abuse, not just the existence of a market monopoly.

ALCOA had been a very friendly monopoly and its paternal approach towards workers had garnered praise. It maintained a reasonable price and one that was often below that of international producers. ALCOA controlled 100 percent of primary ingot production, and in fabricating, it controlled 60 percent of sheet, 60 percent of extrusions, 40 percent of forgings, 60 percent of tubing and wire, 50 percent of foil, 50 percent of powdered aluminum, 50 percent of utensils, and 22 percent of cast pistons. The biggest barrier to entry for others was ALCOA's exceptional expertise, especially in the processing of aircraft alloys. This expertise gave ALCOA a huge advantage and made it an integral part of the defense industry. ALCOA was considered an American company even though it had interests in several international companies. Specifically, ALOCA owned foreign mines which controlled the price of bauxite.

The other question under the Sherman Act was the participation in foreign cartels. Here again, it would be difficult to prove, thanks to the spinoff of Alcan (Aluminium Limited) in 1928. Still, Alcan had the same stockholders as ALCOA; and, of course, there were the family ties of Edward and Arthur Davis. Clearly, Alcan was free to participate in cartels under European law, but the government argued that the purpose of Alcan's formation was to achieve a loophole. In reality, there was no doubt of Alcan's having entered a cartel. In 1931, Alcan (Aluminium Limited) had entered an agreement with British Aluminium, L'Aluminium Française (French), Aluminium Industrie Aktiengesellschaft Neuhausen (Swiss), and Vereinigte Aluminium Werke (German) to form the Foundation Agreement also known as Alliance Aluminium Compagnie. The cartel was indeed the creation of Edward Davis, but did not include ALCOA. Its main purpose was to control prices in the oversupply of the depression; and as a Canadian company, Aluminium Limited had every right to do so. The government tried to prove ALCOA was in a shadow agreement through Alcan and, therefore, part of an illegal cartel.

In 1937, the Justice Department of the Roosevelt administration charged ALCOA with illegal monopolization under the Sherman Act and demanded that the company be dissolved. ALCOA's participation in an alleged bauxite cartel and its ties to Aluminium Limited and European cartels allowed for 90 percent control of aluminum ingot in the United States. Additional charges alleged that ALCOA had used its control to force competition out of the market. The Roosevelt administration threw the kitchen sink at them with a total of 113 separate charges, hoping something would stick. The Justice Department was well aware of the decades of scrutiny that had resulted in no proof of illegal, or even inappropriate, behavior by ALCOA. The anti-business environment of the late 1930s made it the perfect time to try once again and the scope of the trial was enormous. One historian noted: "The charges and trial embraced virtually the entire history of the aluminum industry in North America, all phases of its operation and sales, as well as ALCOA's acquisition of the European companies which Alcan had taken over."[5] Clearly, the mountains of documents from the trial are a written history of the industry.

The presiding judge for the 1937 trail was Francis G. Caffey (1868–1951) of the U.S. District Court for the Southern District of New York. Caffey was nominated by President Herbert Hoover on April 18, 1929, to a new federal judgeship and confirmed by the United States Senate on April 29, 1929. This trial would become the longest antitrust trial in U.S. history with 6½ months of trial days spread over nearly 5 years, with 58,000 pages of written records. There were 153 witnesses called and Arthur Davis was in the witness chair for 7 weeks. It would take 13 years to establish the final results after appeals.

Arthur V. Davis, president and chairman of ALCOA from the 1920s through 1945 (Library of Congress).

The *New Yorker* put the enormousness of the trial in perspective for its readers in 1942: "The transcript of the testimony weighed 300 and 25 pounds, or more than 3 times as much as the *Encyclopaedia Britannica*. The record, exclusive of the judge's decision, which has been rendered impromptu from the bench but has not been edited into final shape, was printed in 400 and 80 volumes. It contained 15 million words or more than 30 times as many as *Gone with the Wind*.... If the period of preparation is included, the ALCOA case outlasted the Civil War."[6]

In 1937, as the economy took a second dip, the trial became a crusade against the bankruptcy of capitalism for New Deal democrats and populists. New Dealers needed something to take the focus off the administration's poor results, and they saw ALCOA and the Mellon family, once again, as the perfect villain. The 1937 recession has been called the recession within the depression, or a double-dip recession, while others at the time referred to it as "Roosevelt's Recession." By 1937, in the Great Depression, all economic measures were back to pre-depression levels, with the exception of unemployment. Of course, for most Americans, the only indicator that counts is unemployment. Unemployment increased from 5 million to almost 11 million by May of 1938. The economic decline during the years of 1937–1938 was steep and it shook the New Deal policies. The decline in economic activity was greater than that of the years 1929–1933 at the start of the depression. Also, the industrial index fell 37 percent for the period. Roosevelt had lost much political capital in 1937 and 1938, and no New Deal legislation was passed. However, Roosevelt managed to spend another five billion dollars in the spring of 1938, which produced no visible results. The hope was that this great trial would be viewed as dealing with a scandal in capitalism and would capture headlines. It was also a warning of what might happen to other businesses that opposed unionization.

The size of the investigation seemed to expand every day, and the trial appeared to have an endless line of company officers of ALCOA and Alcan who were brought in or placed on standby as witnesses. Andrew Mellon, one of the defendants, would die at age 82 on August 28, 1937, as the trial was underway. Many actually viewed the loss of Mellon as a benefit to the company because of his public reputation. It wasn't until the 1950s that the trial was fully resolved and the company responded. As war approached, public concerns began to arise about breaking up the strategic aluminum industry of the United States. Even some in the Roosevelt administration doubted the wisdom of breaking up America's prime defense industry.

The initial trial concluded by finding no fault in ALCOA to damage other producers through its monopoly. The government argument on cartel participation centered on at least a tacit understanding between ALCOA and Aluminium Limited and had made ALCOA party to Alliance Aluminium Compagnie. In the end, the court did not find a tacit relationship. Furthermore, the Alliance's control of market only applied in Europe and did not address the

American market. The court also rejected the accusation that Aluminium Limited was formed as a device to loophole American antitrust laws. ALCOA was cleared of all other minor charges of direct involvement in foreign cartels.

ALCOA would also be cleared of other charges against them. One major concern for ALCOA was the government contention of illegal conduct in domestic competition. Certainly, there was substantial evidence that ALCOA did govern the sheet market by controlling prices. Independent sheet fabricators had to buy ingot at set prices. The problem came with the testimony of George Haskell, considered the major victim in the case, who stated ALCOA had not intentionally tried to force Brush Machine Tool out of the business, but rather the sheer size of ALCOA restricted his ability to compete. Again, the argument became monopolistic size versus intent. The government, feeling a setback in its efforts to prove direct intent, argued that ALCOA's size was by design. For Judge Caffey, ALCOA had committed wrongdoing in trying to expand and grow, and this point would become important in a later appeal. Part of Haskell's sheet challenges were directly related to the technology of rolling of Duralumin, and ALCOA was found to be within its rights not to share the information. The production of Duralumin sheet gave ALCOA full control of the aircraft skin market.

Near the end of the trial Judge Caffey noted: "The astonishing thing is the number of witnesses who appeared on the stand, competitors as well as customers of ALCOA, who have completely exculpated ALCOA from blame and have praised its fairness as well as helpfulness to the aluminum industry. I think I should add that such conduct of those witnesses, nearly all of whom were entirely independent, is in great part a tribute to Mr. Arthur V. Davis."[7]

What was even more amazing was that the court's opinion implied that ALCOA's monopoly control of aluminum was not in question. To even the casual observer and many in ALCOA, it represented a monopoly, but Judge Caffey summarized at the end of his opinion with the following: "I think it is clear that, with the access to the two raw materials of ore and power named which is and, save when prevented by a patent, always has been open to everybody in the United States, anyone possessing the four cardinal tangible elements of intelligence, industry, courage, and money or credit is and has been able, with confidence, to go into the production of virgin aluminum. Anyone in the United States outfitted with the four prerequisites I have mentioned is now free, and since the expiration of the Bradley patent in 1909 has been free to produce virgin aluminum."[8]

Perhaps the important point made from a business standpoint was Judge Caffey's argument that finding ALCOA guilty would be to penalize excellence in operations and individual effort. Judge Caffey believed that both capitalism and American innovation were on trial. He believed that companies should pursue bigness, and in fact, it was their obligation to their stockholders to do so. Caffey found no evidence for any of the charges brought by the government, and he went further by setting ALCOA as an ideal for capitalism and American business. Such a decision did not sit well with big government New Dealers of the Democrat Party, who saw big business as the problem.

ALCOA won the trial on all 130 counts, but astonishingly, the government won their appeal in 1942. Even though ALCOA was a monopoly, it showed no intent to monopolize or efforts to control the market. The long trial, as well as previous trials, had bound up the judiciary. Review by the Supreme Court was impossible, since four of the justices had been involved in prior antitrust suits against ALCOA. A special act of Congress was necessary to give the 2nd Circuit Court of Appeals the authority of a Supreme Court opinion in the appeal. The 2nd Circuit Court upheld the original decision on all of the 130 counts except one. The court found that ALCOA did control over 90 percent of the U.S. market for aluminum ingot; and, therefore, this represented monopoly control over ingot production. The Appeals Court ruled that this 90 percent control of ingot was alone sufficient to show violation of the Sherman Act, regardless of the intent to monopolize. Judge Hand wrote the opinion for the 2nd Circuit Court, ruling that ALCOA's claim stating its ingot competed with scrap was erroneous, and that the company was a monopoly in primary aluminum.

Prior to the ALCOA case, the judiciary had focused on antitrust law based upon whether the company was a "good" or "bad" monopoly. The decision would change antitrust law, which had required wrongdoing. Judge Hand ruled it did not matter how ALCOA became a monopoly, since its offense was simply to become one. The ruling caused further confusion as Judge Hand acknowledged it might be different if ALCOA had not tried to become a monopoly. If ALCOA or any company just became a monopoly, without any corporate focus, then there would be no wrongdoing, no liability, and no need to remedy the situation. The judge's acknowledgment has generally been considered as an empty one in the context of the rest of the opinion.

The court recommended dissolution, but the war would intervene and implementation was placed on hold until 1945. The main result was that major stockholders of ALCOA had to divest in Alcan stock. While Alcan was divested, the rest of the court orders had to wait until after the war. In the meantime, the Roosevelt Administration continued to set up ALCOA for a breakup following the war. Full remedy required the wartime Surplus Property Board to sell off wartime assets to Kaiser and Reynolds, creating oligopoly in primary production in 1945. ALCOA was required to license critical alumina production technology to competitors, royalty-free. Aluminium Limited (now Alcan) had to effectively spin off from ALCOA by having joint stockholders divest in 1942. The court struggle was still not over until the 1950s as the war and politics continued to cause distractions.

A secondary result of the court struggle was a change in the focus of management that took place after the war. In particular, the once highly regarded research function of ALCOA was forced from long-term improvements to short-term processes and cost-control development; and ultimately, ALCOA became less interested in developing new uses for aluminum. In 1948, ALCOA asked the court to reconsider the decision and, once again, the decision was reversed, finding ALCOA guilt-free on all charges. However, it would be too late to change the previous results. In 1947, ALCOA made the argument to the court that there were two effective new entrants into the aluminum market, Reynolds and Kaiser, as a result of the government's divestiture of defense plants. In other words, the problem had solved itself and no judicial action would be required. On this basis, the district court judge ruled against divestiture in 1950, but the court retained jurisdiction over the case for five years.

Many have argued that the approach of the court and the Roosevelt administration to break up efficient companies was an anti-capitalist, and even un–American, decision. Former Federal Reserve Chairman Alan Greenspan criticized *United States v. ALCOA* in 1966, in an essay published in *Capitalism: The Unknown Ideal*. In the essay, he argues that antitrust law should only condemn coercive monopolies: "ALCOA is being condemned for being too successful, too efficient, and too good a competitor. Whatever damage the antitrust laws may have done to our economy, whatever distortions of the structure of the nation's capital they may have created, these are less disastrous than the fact that the effective purpose, the hidden intent, and the actual practice of the antitrust laws in the United States have led to the condemnation of the productive and efficient members of our society because they are productive and efficient."[9] The debate on monopolies and what is illegal continues to this day, as with the Microsoft case.

So how might a business professor or an average business student assess the court decisions today? There would be little doubt that ALCOA was a monopoly and had a market advantage. While Judge Caffey believed that a driven capitalist could break into ingot production, he did note one of the requirements needed would be money. Caffey was right, with the exception of the amount of money to secure energy and raw materials, which would be a substantial barrier to entry. Henry Ford, for example, might have that kind of money, but very few others would be capable of producing those significant funds. ALCOA's position was achieved fairly and competitively and most likely used that advantage on a tactical level. This behavior, however, is a tenet of capitalism.

The argument that aluminum required a new type of capitalism does make sense, but capitalism was fundamental to the competitiveness of ALCOA. Clearly, size alone should not be the basis of breaking up a company, but instead, all should be viewed when looking at the impact on the American market alone. Most average citizens would agree that the criteria should focus on wrongdoing, not size alone. In fact, in today's globalized market, American case law from the ALCOA trials puts American companies at a disadvantage.

Currently, American monopolies are needed to compete internationally against government-supported monopolies and cartels of China, Japan, Russia, and Europe. The idea that ALCOA and Aluminium Limited did not have some tacit arrangements does stretch the imagination, since the companies shared major stockholders and both were run by brothers. The ALCOA and Alcan stock arrangement was clearly a trust like that of Standard Oil in 1911. Again, one must consider if this is in itself illegal or damaging to the American marketplace. Some could argue such an arrangement was to the advantage of the American consumer and, in the case of war, a plus to the nation. In the end, even with the loss on appeals, ALCOA seems to have received a fair deal through the courts. Many would ask whether breaking up companies such as Microsoft and Apple should be judged on impact to American citizens. Even with Microsoft being found guilty of monopolistic behavior in the 2000s, the Justice Department decided not to break up the company, and rightly so.

CHAPTER 13

The War and the New Aluminum Industry

WORLD WAR II WOULD CHANGE EVERYTHING for ALCOA and the aluminum industry. As big as ALCOA had been, it did not have the capacity for war. In late 1938, during the antitrust trial, ALCOA started to build inventory as the war in Europe threatened to expand to America. The war would offer its own solution for both the government and ALCOA by allowing ALCOA to expand while the government entered into the marketplace. Eventually, two new aluminum companies, Reynolds and Kaiser, would enter the aluminum business as the market for aluminum mushroomed. In World War I, U.S. military requirements for aluminum were 128,867 tons, whereas in World War II, America used 1,537,590 tons. The United States would build 304,000 aluminum planes, which had 90 percent aluminum wings and fuselage, 60 percent aluminum in the engines, all-aluminum propellers, and hundreds of smaller aluminum parts and instruments, for a total of 3.5 billion pounds. In addition, aside from airplanes, a single tank required 18 tons of metal, and just one of the navy ships comprised as much as 900 tons. In 1939, the ranking of aluminum production was led by Germany and followed by the United States, Canada, France, Russia, and Switzerland.

Hitler's Third Reich issued coins made from aluminum. Starting in 1939, a swastika appeared on the eagle's feet on the 50 Reichspfennig. This time, the German aluminum coin was not a result of inflation and decline, but a symbol of growth and power. The Waffen SS even produced an aluminum belt buckle. Germany was by far the most creative in terms of using aluminum in war. Special aluminum knives made of Duralumin were issued to pilots and gilder troops. Aluminum was used in some helmets, and badges for officers' hats, such as the German eagle, were cast aluminum. Also, special aluminum tableware was produced for German officers. The Nazis saw aluminum as their edge in the war, since Germany was the number one producer. Overall, aluminum production and alloy development by the Germans were unequalled. Norway was an early target of Nazi Germany because of its cheap energy; and soon after the Nazis invaded Norway, joint aluminum ventures were formed to increase aluminum production. Norway never produced the aluminum the Reich had planned for because the Allies effectively blocked bauxite shipments. ALCOA, however, proved up to the task of matching Germany during the war.

A 1946 *Fortune* article summarized the amazing contribution of ALCOA to industry and the war: "Altogether, between the start of the defense program in May, 1940 and V-J Day, ALCOA produced 11.4 billion pounds of alumina, smelted 5.5 billion pounds of aluminum, and fabricated 2.7 billion pounds of extruded shapes, 500 million pounds of forgings, and 400 million pounds of castings. Aluminum and ALCOA may not be synonymous as they once were, but if the U.S. won the war, they, too, may be said to have won it."[1]

ALCOA's war contribution was greater than that of any other company. ALCOA invested over 450 million dollars of its own money to expand production. ALCOA shut down all domestic operations, such as kitchen utensils, to supply the military. Another 473 million dollars was given to ALCOA from the government to build new "government" plants, and an additional 100 million went to Alcan to expand in Canada. The plants were the property of the Defense Plant Corporation, but ALCOA was in charge of site selection, construction, and operation. These "government" plants were leased to ALCOA and, under the contract, 85 percent of the profit from these plants went to the government. ALCOA, on its own, invested additional millions in these plants. ALCOA contracted and ran for the government 8 smelters, 4 alumina plants, and 11 fabricating plants. With the Germans being the world's largest producer, the Defense Department wanted the knowledge and resources of ALCOA where most money was directed. New Dealers and the Roosevelt administration did try to address ALCOA's monopoly by getting Reynolds Metal, the major aluminum foil producer, into the production of alumina and primary aluminum. They loaned Reynolds the money to build a new plant that had a 40-million-pound capacity in Listerhill, Alabama, in the Muscle Shoals district.

ALCOA proved to be very flexible and goal-oriented throughout the war, but Roosevelt was continually pushing his antitrust case against ALCOA and watching the government contracts with ALCOA. Roosevelt's czar of war production, Jesse Jones, saw things differently. ALCOA was one of America's best-run companies, capable of meeting and exceeding all government requirements. President Roosevelt, seeing a newspaper article on the building of new ALCOA plants, sent Jones a handwritten note: "What about this story about 12 new aluminum plants—all to be operated by ALCOA?" Jones's reply was: "We find that they are well organized and progressing on schedule."[2] Jones also made it publicly known that he had included Reynolds in some contracts for appearances of competition. ALCOA was truly integral to the defense industry, whether Roosevelt liked it or not.

Alcan's contribution to the war in Canada was just as important. In the 1930s, after the split from ALCOA, Alcan struggled to develop its own markets, but the war opened world markets. Alcan became the second largest supplier of aluminum and supplied 90 percent of the British needs and 35 percent of the Allies'. Regardless of the court action, Alcan and ALCOA worked closely and cooperatively on technical issues and natural resources.

The problem for Alcan and ALCOA was a threat of the Germans cutting off cryolite supplies. Cryolite, as we have seen, was the key to the unleashing of aluminum from its ore. By the late 1930s, most companies had learned to use a mix of natural cryolite and artificial cryolite, but cryolite was still needed for efficiency. Greenland was the only supplier of cryolite for aluminum companies.

In 1940, Greenland was a Danish colony. The fall of Denmark to German invasion in April 1940 left Greenland in an uncertain position. Cryolite had been stockpiled by Germany, which also made artificial cryolite. In 1940, the United States remained neutral and declared Greenland was neutral, preventing the British from invading it. Although the Danish government continued in power and still considered itself neutral, it was forced to obey German wishes in foreign policy matters. The most immediate concern for the United States was that Germany would attack the cryolite mines. In 1940, the chief concern of all interested parties was to secure the strategically important supply of cryolite from the mine at Ivigtut. Due to diplomatic considerations, no U.S. soldiers could be used to protect the mines, so the U.S. State Department recruited 15 Coast Guardsmen who were voluntarily discharged and in turn hired by the mine. In addition, three-inch naval guns were supplied, along with eight machine guns, fifty rifles, and thousands of rounds of ammunition. In this way the United States maintained neutrality and still preempted British and Canadian plans for the island. When the United States entered the war in 1942, the U.S. Army took over protection of the Ivigtut mine, and combat air patrols flew over daily. The Germans only made some minor incursions.

While the threat never materialized, it did cause a panic for Canadian Alcan and ALCOA. Both companies sped the research and development efforts in artificial cryolite. This research had already started in the 1930s as the cryolite ore had continued for some years to decline in quality. Alcan took the lead in using enriched fluorspar from Newfoundland. In this respect, the war was again

13. The War and the New Aluminum Industry 137

DC-3, the first commercial aircraft using an aluminum body (Library of Congress).

fortuitous in that by 1950 most of the Greenland mines were exhausted, and the industry could run on artificial cryolite.

The real fear concerning war production was a possible interruption in the supply of alumina ore. British Guiana was a major supplier of high-grade bauxite to the United States during the war years. Guiana was a British colony on the northern coast of South America, since 1966 known as the independent nation of Guyana. The aluminum produced from this bauxite was used by the American military for the production of aircraft. Guiana accounted for over 60 percent of the supply of bauxite to the United States and Canada. Significantly, roughly two-thirds of all Allied aircraft manufactured during the war years used aluminum made from Guianese bauxite. As a result of the booming demand for Guiana's bauxite, exports increased from 476,000 tons in 1939 to 1,902,000 tons in 1943. Even so, many bauxite shipments were lost to German submarines. At the peak year of 1942, bauxite shipments were cut off for weeks at a time. Well over 50 ships were torpedoed during the war.

Germany was concentrated on the destruction of aluminum manufacture in America, seeing aluminum as necessary to victory. ALCOA responded to the bauxite crisis with the development of a new refining process that could utilize lower-grade American bauxite. This was the patented ALCOA Combination Process, which was the first major breakthrough in refining in decades. The process allowed ALCOA to glean low-grade ore from its Southern mines. The process, however, required three tons instead of two to produce a ton of alumina. Charles Hall had predicted that clay would be a future ore, but cost was the major roadblock.

The government pushed for research into "non-bauxite" ore. In addition, processes for recycling red mud waste were developed to glean out even more alumina. In 1942, the federal government, through the Metals Reserve Company, undertook to build a large stockpile of lower-grade Arkansas bauxite as insurance against the menace of German submarines. Aluminum was so important that in 1942, eight German saboteurs landed from U-boats, four on Long Island and four just south of Jacksonville, Florida, on a mission to destroy ALCOA's plants in Alcoa, Tennessee, Massena, and East St. Louis. The German saboteurs had been trained in German aluminum plants to understand how to destroy production processes.

The war also opened new exploration for bauxite as well. South America would become the main supplier of alumina from a Dutch colony, Surinam. Surinam, today an independent state on the northeastern Atlantic coast of South America, is bordered by French Guiana to the east, Guyana to the west, and Brazil to the south. Surinam was believed to have the most reserves of bauxite in 1940. ALCOA had done some mining here going back to 1917. On November 23, 1941, under an agreement with the Netherlands' government-in-exile, the United States occupied Surinam to protect the bauxite mines. By 1942, Surinam was functioning as an American colony with the occupation of the U.S. army to protect the mining. Many criticized the protective ownership of bauxite mines of Surinam as stealing resources from the indigenous people. Because of shipping shortages in World War I, ALCOA developed its own shipping line to carry bauxite from its source in Surinam and Guyana to aluminum mills in the United States and elsewhere. At first the line operated under foreign flags. From 1940 it operated under the U.S. flag for full protection.

In 1942, a wealthy landowner on the Caribbean of Jamaica was having trouble growing grass on his red soil. Analysis of the soil showed he was sitting on one of the world's major deposits of bauxite. Jamaican bauxite was not used during the war, but three North American companies (Alcan, Kaiser and Reynolds) came to the island to survey, acquire reserve lands, and set up operations. Reynolds began exporting bauxite from Ocho Rios in June 1952, and Kaiser followed a year later from Port Kaiser on the south coast. Alcan built the first alumina processing plant near its mines at Kirkvine, Manchester, and in early 1952 began shipping alumina from Port Esquivel. This was the beginning of the alumina industry in Jamaica that would dominate in the future. The protection and discovery of ore was a critical first step for the military. Technology, however, was just as important.

Aluminum airplanes were the key to victory in the war, and that would require major steps in construction and alloy design. Japan and Russia proved to be ahead of the Americans and even the Germans in aircraft design. Japan was by far the most impressive, having only started in 1933. Using a tariff of 20 percent on incoming aluminum in the 1930s, Japan built up its home aluminum industry. Japan experimented with different ores to assure production and built its research operations. The result was the development of an array of new Japanese aircraft alloys. America needed to develop new materials for its aircraft if they were to compete with the Germans and Japanese. It must be remembered that the Germans and Japanese had been preparing for war 10 years ahead of the Americans. Both countries' governments had taken over all metallurgical research in their respective nations to focus on war needs. In the United States, ALCOA alone had the resources; but during the 1930s, the total focus was not war; actually, it was far from it.

ALCOA did an amazing job of catching up in aluminum alloys. Using ALCOA resources alone, ALCOA's technology quickly approached years of work done at government-based research centers in Europe, Russia, and Japan. Aluminum research played an important role during World War II. Russia, like America, stepped up to the challenge. The invaluable contribution in establishing the defense power of the Soviet Army was made by the Urals Aluminum Smelter (UAZ). The first stage of UAZ was commissioned in September 1939. On the eve of the war, 36 percent of aluminum produced in Russia was produced at UAZ. A Russian conversion of high-strength Duralumin sheets and slabs served as the main material for airplane covering. Complex pre-formed blocks were produced and then forged into component parts of airplane engines, propellers, the chassis, and the fuselage frame. Soft low-alloy Duralumin and aluminum-magnesium alloys were used for rolling wire for rivets, covering connective elements; sheets of aluminum-manganese alloy were used for welding fuel tanks.

Today we have the following standardized series of alloys, which evolved out of war research. The 1000 series classification is reserved for alloys of nearly pure aluminum metal. They tend to be less strong than other alloys of aluminum. These metals are used in the structural parts of buildings, as decorative trim, in chemical equipment, and as heat reflectors. The 2000 series are alloys of copper and aluminum. They are very strong and are corrosion resistant. This series was born out of the German Duralumin. Some applications of 2000 series aluminum alloys are in truck paneling and structural parts of aircraft.

The 3000 series is made up of alloys of aluminum and manganese. These alloys are not as strong as the 2000 series, but they have good machinability. Alloys in this series are used for cooking utensils, aluminum furniture, highway signs, and roofing. Alloys in the 4000 series contain silicon. They have low melting points and are used to make solders. The 5000, 6000, and 7000 series include alloys consisting of magnesium, manganese and silicon, and zinc, respectively. These are used in boat production, structural parts in buildings, automobile parts, and aircraft components. It would be World War II that inspired the development of these alloys.

The United States quickly closed the technology gap when the war came to America in the 1940s, but work on alloys had been evolving in the 1930s as well. First, ALCOA was able to improve some of its earlier alloys, such as 17S, successfully used in commercial aircraft. In 1939, the U.S. produced 148,000 tons of ALCOA's 17S sheet alloy. It was based on the German alloy Duralumin, and it had been used to build the first commercial all-metal passenger airplane in the U.S., the Ford Trimotor in 1929. The Ford Trimotor was an aluminum plane, and it pioneered aluminum in the aircraft industry, borrowing from Germany's use of aluminum in the Junkers. The Trimotor was primarily designed as a passenger aircraft, capable of flying up to 10 passengers, and was used by over 100 newly formed airline companies, such as Transcontinental Air Transport (later to become TWA), and Pan American Airways (Pan Am). The Trimotor used corrugated aluminum in the wings for strength. Ford also used aluminum chairs in the plane. Other passenger planes in the 1930s were wood, but a famous air crash ended wood's role in passenger planes. On March 31, 1931, Knute Rockne, the famous Notre Dame football coach, was killed when a wooden Fokker trimotor crashed and burned. It had suffered a structural failure partly because of its wood construction. Consequently, the Civil Aeronautics Authority grounded the plane and insisted on so many modifications that the Fokker was taken out of service in America, leaving the company to return to solely European production.

Aluminum offered a solution, but stronger alloys were needed. The Ford Trimotor was eclipsed in 1934 by the next generation of all-metal smooth-winged aircraft, such as the Boeing 247 and Douglas DC-2, which used these new alloys such as 17S. These planes could handle 12 to 14 passengers and travel at faster speeds. ALCOA was able to use the 17S alloy for some of the first military aluminum planes in the 1930s as well. The U.S. Army Air Corps' first all-metal monoplane bombers were the Boeing B-9 and the Martin B-10. Both were produced during 1932–1933; the B-9 was outclassed by its contemporary all-metal Martin B-10, and only seven were purchased. The Air Corps' first all-metal fighter was the Consolidated P-25 of 1933. As war approached, much better metal planes would be needed. This new generation of planes required much stronger aluminum alloy than 17S.

ALCOA research developed new heat-treatable stronger alloys for advances in aviation. In the 1930s, ALCOA developed a higher-strength alloy called 24S. The major change from 17S to 24S involved boosting the magnesium level from 0.5 percent to 1.5 percent. The new alloy was used to construct the first commercially successful passenger plane, the Douglas DC-3, in 1935. The DC-3's heavier payload and faster speed made it the most popular commercial airliner of the decade. Alloy 24S was basic aluminum alloy used for nearly all of the 300,000 planes built in the U.S. during World War II. Still, the bigger bombers for war required even more strength and corrosion resistance. These new alloys in the 6000 series would also become the foundation of America's future space program.

Another heat-hardening alloy system developed by ALCOA in the 1930s added 1 percent magnesium, 0.6 percent silicon, and 0.3 percent copper to aluminum. This alloy was called 61S (now 6061) and would become the iconic aluminum alloy. The 61S found immediate applications in shipbuilding to reduce weight, and would be key to warplanes. It is the structural material for a great tonnage of ordinary engineering applications even today. The new Ford F150 aluminum truck today uses 6000 series aluminum. A number of alloys based on 6061 were developed during the war, containing additional alloying elements, and are still widely available as well. These aluminum alloys are known for ease of fabrication, corrosion resistance, and low cost compared to other high-strength aircraft alloys. The alloy 6065 was developed for rivets for airplane skins. The 6065 rivets were worked hot, and precipitation hardening created a tight fit on cooling. These alloys were and are used for general industrial applications such as trucks, buses, rail cars, trailer tanks, storage tanks, building construction, and light aircraft. Even today, 6061 is the most common manufacturing alloy in the world. These alloys required a coating of pure aluminum to prevent saltwater corrosion, a technique ALCOA perfected.

Alloy 6061 would be the alloy selected to represent the technology of man decades later. It was used to make an engraved plate to the *Pioneer 10* and *11* space probes in 1972. The first plaque was launched with *Pioneer 10* on March 2, 1972, and the second followed with *Pioneer 11* on April 5, 1973. The *Pioneer 10* and *11* spacecraft were the first human-built objects to achieve escape velocity from the Solar System. The plaques were attached to each spacecraft's antenna support struts in a position that would shield them from erosion by stellar dust. The idea that the *Pioneer* spacecraft should carry a message from mankind to extraterrestrial life was that of the famous astronomer, Carl Sagan. NASA agreed to the plan and gave him three weeks to prepare a message.

Together with Frank Drake he designed the plaque, and the artwork was prepared by Sagan's then-wife, Linda Salzman Sagan. The plaque was gold-color anodized 6061. While 6061 may have been the iconic structural alloy of the 1930s, that decade would bring a series of futuristic alloys that changed the industry forever.

Even stronger alloys and corrosion restraint alloys using zinc and chromium were developed for specific applications during the war. Aluminum alloys that contain zinc, magnesium, and copper were originally studied in Germany, but ALCOA was quick to improve on them. Alloys featuring zinc as a major alloying element exhibit very high strengths. The problem was a tendency to stress-crack under corrosive environments. Research on these alloys was performed at ALCOA, and the first commercial composition was 76S, used for aircraft propellers in 1940. ALCOA was able to address stress corrosion cracking by adding small amounts of chromium to the alloy. This led to the commercial alloy 75S (now 7075), which contains 5.5 percent zinc. This innovative alloy was introduced during World War II as the structural metal on Boeing's B-29 Superfortress long-range bomber. The reduced weight due to the 75S enabled the addition of significant numbers of bombs to the B-29 payload. Most of the Japanese aircraft were built on their top-secret equivalent of 75S (7075) aluminum alloy developed by Sumitomo Metal Industries in 1936, ahead of ALCOA's research. ALCOA was probably aided by the capture of a Japanese Zero early in the war. In June of 1942 a Zero crash-landed but was virtually undamaged. The Zero was recovered in the Aleutian Islands and the U.S. military was able to see and fly, firsthand, this advanced fighter aircraft, as well as analyze the aluminum used. Amazingly, this alloy 7075 today is key in military applications such as the M16 rifle; industrial uses such as aluminum tubes to enrich uranium; and consumer applications such as the I-watch.

Where ALCOA had the lead was in the coating of 75S alloys, which the company started in the 1930s. Unlike steel, aluminum alloys do not exhibit signs of rapid and visible oxidation. Surface oxidation is still present and shows as a matte grayish coating; but as soon as it is rubbed and properly cleaned, the aluminum becomes again smooth and shiny. This was particularly true in saltwater environments. Nonetheless, it was discovered that aluminum could be affected by a more dangerous internal form of oxidation, called inter-granular corrosion, in the form of microscopic black spots which literally eat the metal and make it fragile and brittle. The cause was not the aluminum itself but rather impurities in the metal. Again, salt water accelerated the corrosion. To combat this problem, the aluminum alloys were coated by an almost 90 percent pure aluminum called "Alclad." This technique was known (probably from ALCOA) to the Japanese, who used it extensively. By the end of the war, America was far ahead in aluminum aircraft technology. American metallurgists were unsurpassed in all phases of aluminum applications.

One of the strangest cases of aluminum metallurgy in World War II was in unconfirmed reports of efforts to sabotage German planes. Aluminum is anodized to protect aircraft skins. This invisible barrier forms so quickly that aluminum seems to be an inert metal. But this illusion can be shattered with aluminum's one enemy, mercury. Mercury can cause aluminum to crack or rust. A *Popular Science* reporter noted: "I've heard that during World War II, commandos were sent deep into German territory to smear mercury paste on aircraft to make them inexplicably fall apart. Whether the story is true or not, the sabotage would have worked."[3] While this rumor is an old one, it does lack citable proof. ALCOA's research and American manufacturing, more importantly, led to the development of the world's greatest air force.

The American military aircraft of World War II was a monoplane with one to four engines and an aluminum airframe housing a mass of equipment for the purposes of navigation, armament, communication, and crew accommodation. The engine and its accompanying aluminum propeller were the keys to aircraft performance for speed, range, altitude, and rate of climb, and depended in large measure on the power. Aluminum production and aircraft manufacture output of American factories were twice those of all the Axis nations as manufacturers everywhere shut down normal operations. The quantity of aluminum needed for this vast undertaking greatly exceeded the capacity of ALCOA, the only aluminum manufacturer in the country. By 1943, when wartime production reached its peak, the U.S. produced 835,000 tons against 250,000 tons in Germany. In addition, Canadian production increased from 75,000 tons in 1939 to 450,000 tons in 1943. Overall production for the 5 war years reached 4 million tons in the U.S. and Canada versus 1.4 million tons in Germany.

The conversion of America's auto assembly lines to aircraft production was a huge advantage over the German and Japanese crafts-type assembly. Americans were making planes at a rate ten times faster than Germany and Japan. Both countries were building airplanes like they did large machines. At the time, neither Germany nor Japan processed any prewar assembly lines. The Ford Motor Company proved adept at converting auto assembly lines to the fabrication of airplanes and tanks. America also pioneered new assembly technology such as inert gas welding of aluminum.

Another ancillary advantage was America's vast hydroelectric resources. Again, these resources far exceeded those of our enemies. In 1943, the aluminum industry was the largest single electricity user in the U.S., consuming 22 billion kilowatts annually. By 1945, the company's power plants in East

Tennessee and Western North Carolina were furnishing 50 percent of ALCOA's power, with the other 50 percent purchased from the Tennessee Valley Authority. The TVA of the New Deal 1930s proved significant to aluminum production. To provide power for such critical war industries, the TVA engaged in one of the largest hydropower construction programs ever undertaken in the United States. Early in 1942, when the effort reached its peak, 12 hydroelectric projects were under construction.

During the war, ALCOA managed to maintain its aluminum monopoly with the tacit approval of the government. In fact, the military had grown dependent on ALCOA for its expertise. With the granting of loans for Reynolds Aluminum to expand into ingot with the help of the Roosevelt administration, ALCOA was threatened. The company began diverting more aluminum to the military, cutting shipments to Reynolds. Reynolds was forced to work off inventory to keep running. The Roosevelt administration would make this a key point in its antitrust case after the war.

The end of World War II would once again bring calls from the Democrat Party to break up the ALCOA monopoly. Harry Truman's administration would prove to be, like Roosevelt's, in favor of the breakup of ALCOA. Truman had been part of the overseeing of the war production and had questioned the strong link with ALCOA. Truman's Attorney General Thomas Clark argued for the breakup of ALCOA to satisfy the antitrust verdict of 1945. Truman, however, lacked the political capital of the early Roosevelt administration. The legal clash between the government and ALCOA over the verdict continued, and in 1945 the adversaries reached an impasse. The government tried to rearrange the aluminum industry through the sale of its wartime plants. The Defense Plant Corporation actually controlled 52 percent of the nation's aluminum, with ALCOA having 38 percent and Reynolds 7 percent in 1945. The government decided to go public with the sale of aluminum plants.

Arthur Davis was extremely angered at the government's hardball policies. Davis fully believed ALCOA deserved better after its war efforts. The Defense Plant Corporation, however, found the sale of these plants tough going. The reason was ALCOA controlled the expertise and owned the technology. ALCOA was able to buy these resources and put up a major barrier to entry for competition. The government tried to have Reynolds take some of the plants since they had processing expertise. ALCOA, however, still controlled much of the processing through patents. The Truman administration countered with pressure on Congress to force an internal breakup and restructuring of ALCOA.[4] The company had gained some public support during the war, which did not make such legislation a sure thing.

Neither side wanted more endless legal battles, so negotiations went forward on a compromise. Davis moved out of the negotiations, turning them over to I.W. Wilson, executive vice-president of ALCOA. One major roadblock was the sale of the Hurricane Creek smelting plant in Arkansas to Reynolds Aluminum. Reynolds could not hope to efficiently run the plant without the patented ALCOA Combination Process. In addition, there was a huge surplus in aluminum after the war, driving prices down. The negotiations reached an impasse, at which point the government actually suggested that Reynolds buy the plant and operate it, using the ALCOA process, and wait for a patent suit. An agreement was reached with the government and ALCOA, which allowed Reynolds to use up to 25 percent capacity royalty free; and after that point, Reynolds would pay $1 per ton. Reynolds purchased 172 million dollars' worth of aluminum production facilities for $57 million. This included smelters at Hurricane Creek; Jones Mills, Arkansas; and Troutdale, Oregon.

The aluminum profit was so good the new company of Kaiser Aluminum emerged. Kaiser was a major magnesium producer and had light metal experience. The company picked up a major alumina plant at Baton Rouge, Louisiana. Kaiser also got smelters at Spokane and Tacoma, Washington, for the War Surplus Board. In all, Kaiser got 120 million dollars' worth of manufacturing assets for 43 million dollars. By 1950 the rough capacities of these three companies were 50 percent for ALCOA, Reynolds at 38 percent, and Kaiser at 18 percent. The long-desired competition in the aluminum industry had finally been achieved.

The market changes would also result in major internal changes for ALCOA. During the war, Davis, who had ruled ALCOA as an autocrat, turned more responsibility over to Roy Hunt, company president, and I.W. Wilson, executive vice-president. They ruled more like joint emperors, with Hunt handling finance and operations while Wilson focused on technology. Roy Hunt hated organizational charts, running an informally flat organization with decentralized functions. Both Hunt and Wilson had a corporate-level committee to make decisions and coordinate policy. In 1951, Roy Hunt retired as president and Irving W. (known as Chief) Wilson became the first president from a non-founding family. The war had forced a more centralized and structured organization. Army and Navy buyers hated the decentralized and committee-driven decision making at ALCOA during the war, but Roy Hunt had made things happen. ALCOA remained a company, and like Ford Motor Company, it was maintained by a vision. The vision united managers. I.W. Wilson would successfully make the transition from a large family company to a corporation. The Mellon family share of the company dropped to about 28 percent. Things were changing and would have to as ALCOA entered into a new age of global competition. The company entered that market with the hand of the government dividing the market.

Davis was never satisfied with the government deal, but ALCOA would eventually overcome the competition. After the war, Davis had informally turned operations over to I.W. Wilson and formally retired in 1958. Davis died in Miami in 1962 as one of the country's richest industrialists, leaving a $400 million estate. Only a small portion of his wealth went to individuals. The majority went to a trust for Arvida (from *AR*thur *V*ining *DA*vis), the northern Quebec model town he had founded in 1927 for working families. It was proof that Davis was a great paternalist. The Arthur Vining Davis Foundations provide financial assistance to educational, religious, cultural, and scientific institutions. Davis's vision had not died as ALCOA grew from 1948 to 1962 to become the world's largest aluminum company.

CHAPTER 14

Oligopoly and the Dawning of the Golden Age for Aluminum

THE ALUMINUM INDUSTRY AT THE DAWN of the 1950s was in an atmosphere of apprehension. Could the industry survive the switch to peacetime, and how was the overcapacity to be dealt with? ALCOA was no longer in a monopoly position in the United States. The company had Canadian Alcan to worry about on the international ingot market. Alcan, on the other hand, lacked consumer-oriented fabricating and consumer markets. The government had created two major domestic competitors in Kaiser and Reynolds. The path from monopoly to oligopoly offered new challenges for ALCOA. While at the time it looked a bit dark in 1948, some less observed and non-quantifiable factors were at play. No one could fully see the great postwar expansion and prosperity. In the 1950s, aluminum maintained its image as a space-age metal. More importantly, aluminum would put its stamp on American culture in the 1950s. In many ways it achieved all that its advocates had hoped to achieve in the 1930s. As a metal, it started to compete with the venerable steel products. Like cast iron in the late 1800s, aluminum had become symbolic of industry in the mid–1900s.

The postwar surplus would become a foundation for new products. Aluminum had become a known quantity of thousands of manufacturers who would find new applications. The Art Deco movement of the 1930s was morphing into modernism. The Korean War and Cold War would quickly use any surplus and create a shortage. The final result was a wave of investment that raised all aluminum production. In 1945, aircraft producers started to look for opportunities in consumer products. Grumman Aircraft successfully moved into aluminum canoes. Others moved into housing.

Products were developed on the technology of war. One of the biggest uses of aluminum sheet in the late 1940s was aluminum awnings. After seeing how paint bonds to aluminum fuselages on World War II planes, Jerome

Kaufman, owner of one of the largest retail home improvement dealers in the U.S., invented the first residential baked enamel aluminum siding. This allowed for colorful additions to entrances, patios, and windows. The search for new products and uses for aluminum reached the fevered level of the early days of ALCOA.

Transportation was a logical area of increased consumption. The 1930s had seen the dawn of aluminum locomotives, but the war had diverted this development. In 1944, the Louisville and Nashville Railroad placed a rail car order with the American Car and Foundry Company for 28 aluminum, lightweight, streamlined passenger cars. Metal shortages from remaining World War II production delayed completion of the cars, but they were eventually put in service in October of 1946. The new cars were used to create America's first postwar streamlined trains, the *Humming Bird* and the *Georgian*. The cars in the new lightweight fleet were constructed with corrugated aluminum sides. The aluminum train fleet of the Louisville and Nashville Railroad operated into the 1960s, when trains lost out to airlines. Aluminum, however, started showing up in all markets in the late 1940s.

The surprise was that the aluminum global market consumption doubled from 1950 to 1959. ALCOA, Alcan, British Aluminium, Reynolds, and Kaiser all prospered even with the breakup of national monopolies. The stock market reflected this booming aluminum market, and a new generation of aluminum millionaires was created in secondary manufacturers. ALCOA stock went from $46 a share in 1949 to an equivalent of $352 in 1955, Reynolds from $19 to an equivalent of $300, and Kaiser from $22 to $139. The market even had room for three smaller new competitors—Anaconda, Harvey, and Ormet.

All of the smaller three aluminum companies were a result of government stimulation, guaranteed loans, and favorable contracts during the Korean War. The Truman administration, while not obsessed with ALCOA, wanted a highly competitive aluminum industry. These three small companies included Anaconda Copper, which diversified into aluminum production in 1952, when they purchased rights to build an aluminum smelting plant in Columbia Falls. After two years of construction, the plant went online in August 1955. Following two expansions in the 1960s, the plant had a peak output capacity of 180,000 tons annually. Anaconda's experience in metals and finances would make it a long-term competitor in aluminum.

Another of the little three, Harvey Aluminum, was a former aluminum processor and fabricator. Harvey Aluminum had been established in 1914 in Los Angeles and was incorporated in California in 1942. Harvey Aluminum was founded by Leo Harvey, the son of a small factory owner in Lithuania. In 1923 Harvey Machine Company was an aluminum caster and fabricator. During the early '50s, Harvey moved into smelting but was limited by financing. Initially the company had a smelter in Oregon financed by the government. Over the next two decades, Harvey expanded and his inventive powers came to the forefront. Harvey took out numerous patents in specialized machinery and equipment to process aluminum. His patents even included that for the innovative peel-off, pop-top aluminum can.

Ormet came into being in 1955. Ormet Corporation had been in the alumina and aluminum production business when it was first organized by Olin Corporation and Revere Copper and Brass. Ormet built a smelter at Hannibal, Ohio, using coal-based electricity. The company was the smallest fully integrated aluminum producer in the world. It prospered for years until its coal-based energy dragged the company into bankruptcy in 2013. The company only employed 600 at Hannibal, but its power usage is a testimony to aluminum as frozen or packaged electricity. Hannibal used as much energy as the city of Pittsburgh, and it was Ohio's largest energy user. The entrance of Ormet, Anaconda, and Harvey into the smelting of aluminum showed the ALCOA monopoly had been fully broken. The presence these companies also supported ALCOA's legal argument that a small company could enter the business. Still, in the long run, ALCOA had the best control of alumina and energy sources.

ALCOA would not only survive the competition but thrive in it. Its success was rooted not in new strategies, but in the vision of the founders. While the competition in the United States now owned smelters to make ingot, ALCOA's advantage was that it also owned the power to run its smelters. ALCOA, like Hall's small battery problems and Pittsburgh Reduction's shortage of electricity at the first Pittsburgh plant, realized the process and product cost was dependent on power resources. ALCOA had always viewed power as a raw material versus a supplied utility. Kaiser and Reynolds were able, in the short run, to buy cheap power. ALCOA held to the long-term strategy of producing power, even though short-run profits might be increased by purchasing power. ALCOA, as expected, conceded market share and kept to a long-range view. In 1958, ALCOA had 38 percent of the market, 27 percent for Reynolds, 24 percent for Kaiser, and another 11 percent to small producers. Unlike the competition, ALCOA expanded conservatively without taking on debt. Its strategy resulted in high ingot cost in the short run, but secured ALCOA's future.

The big three in the 1950s were ALCOA, Kaiser, and Reynolds. Kaiser Aluminum was founded in 1946 by American industrialist Henry J. Kaiser. Kaiser entered the aluminum business by leasing, then purchasing three government-owned aluminum facilities in Washington State after the war. These were primary reduction plants at Mead and Tacoma, and the rolling mill at Trentwood. The company grew to be a fully vertically integrated aluminum producer. In the 1950s, Kaiser built a plant in West Virginia, which was the first aluminum operation in the world designed as an integrated operation to process the alumina to plate, sheet, and foil at a single site.

Kaiser Aluminum had an end product focus and strategy. Henry Kaiser was a true aluminum visionary, and he helped promote it with the public. In the late 1950s, Kaiser Aluminum, a West Coast company, built an Aluminum Hall of Fame at Disneyland. The exhibit included futuristic uses of aluminum for the space age. Kaiser also envisioned a larger role for aluminum in transportation and inspired a generation of engineers who would take the nation to the moon. Kaiser even teamed up with Disneyland to promote aluminum with kids. On entering Disney's opening of Tomorrowland on July 17, 1955, visitors were welcomed by a friendly pig named Kap. Kap was short for the Kaiser Aluminum Pig. Kap was the mascot for the relatively short-lived "Kaiser Aluminum-Land," the home to the "delightfully told true story of how the sleeping giant of metals—aluminum—was awakened and has become your friendly servant." Kaiser Aluminum-Land, a small sponsorship area, was home to a giant telescope made out of aluminum.[1] Kaiser would build on the consumer market throughout the 1950s and 1960s. Many of those kids would play important roles in the 1960s and 1970s applications of the space-age metal, aluminum. Kaiser also took a serious look at the Dymaxion car and applications in shipbuilding. The company was also on the forefront of promoting aluminum awnings and siding.

Reynolds, of the Big Three, had a long history and had overtaken ALCOA in aluminum foil production in the 1920s. Aluminum foil was considered a strategic war material in World War II. Principal uses during the war were for such essential military applications as packaging to prevent damage to contents by moisture, light, vermin, and heat; electrical capacitors; insulation; and antiradar chaff, which were dropped from planes on bombing missions, and as a radar shield. Reynolds had an edge in the military supply chain as a major defense company and was a natural to vertically integrating into full-scale aluminum production.

Reynolds had also run and managed aluminum smelters for the government during the war. Reynolds looked to achieve competition with ALCOA's low ingot costs through expanding into mining. In 1940, Reynolds Metals began mining bauxite in Arkansas and opened its first aluminum plant near Sheffield, Alabama, the following year. Reynolds also expanded the market through product development. In 1947, the company came up with its Reynolds Wrap Aluminum Foil. Meanwhile, Reynolds Metals pioneered the development of aluminum siding in 1945 and expanded its market rapidly. The government supported reducing Reynolds's ingot cost by making cheap energy available to them. By 1948, Reynolds had the lowest ingot cost in the United States based on its use of purchased energy contracts with the government. ALCOA's ingot cost was even higher than Kaiser, thanks to government support. On an international level, Alcan had significantly lower ingot cost than any of the six American companies. The six U.S. companies, however, acted as an oligopoly; in effect, they controlled aluminum prices and the flow of aluminum to processors in the United States.

The main world competitor after the war for ALCOA would become Alcan. The international picture also created global surplus and new competition for the company. Alcan moved into the war with a lack of fabricating plants, but both the Canadian and British governments would fund their expansion. The British Air Ministry early in 1938 realized the need for Canadian aircraft alloy production. First, it was a supplement to the output of British Aluminium. Second, should German bombing damage production in England, Canada was the backup plan. The war was the perfect time to help Alcan (at the time Northern Aluminium) become independent of its ties to ALCOA.

In 1940, Alcan began production at its sheet rolling and extrusion facility in Kingston, Ontario, with financing from Britain's defense industry. In addition, several smelters were built; and during the same period, Alcan completed construction of a second hydroelectric generating station at Shipshaw, Quebec. Kingston was chosen as the site of the plant partly because it had the advantage of being physically isolated from the war, and the community had a ready supply of available labor. Kingston would become a state-of-the-art supplier of aircraft alloys, both forged and rolled. By 1944 the Kingston Works supplied 40 percent of the aluminum used in Allied war planes, in the form of structural tubes, extruded shapes, forged elements, skins, airframe components, and propellers used in planes such as the Spitfire, Hurricane fighters, the U.S. Superfortress bomber, and the Republic Thunderbolt. In addition, components were sent to the Soviet Union in the latter years of the war. A large research center was also built at Kingston. After the war, the British agreed to let Alcan buy Kingston and other plants, giving badly needed rolling and fabricating operations to Alcan. Alcan after the war had the world's lowest ingot cost.

The war would also give Alcan a major foothold in Europe. In 1939, the British commissioned Alcan (then still Northern Aluminium) to build a large rolling mill at Rogerstone in Wales. During the war, Rogerstone's extrusion and rolling operations were continually expanded. When the war ended, the government was interested in maintaining the jobs created there. The plant switched to prefabricated bungalows known as "AIROH bungalows." AIROH stood for Aircraft Industries Research Organization for Housing. These aluminum houses were constructed from sheet and extruded aluminum from wrecked aircraft and downed German aircraft. Ironically, the Battle of Britain had turned England into a huge aluminum deposit of old German planes. Each AIROH house was built from almost two tons of aluminum, much of which came from German aircraft. A few years later, Alcan purchased the Rogerstone plant; and with further improvements, made it Europe's largest rolling complex in 1950. By the end of the 1950s, Alcan was ready to take on ALCOA in the United States market as well as the British market. Both Alcan and ALCOA would, however, need a new infrastructure for the global competition.

In addition in domestic competition, ALCOA had a structural market problem. While Kaiser and Reynolds were free to expand aggressively into new products and markets, ALCOA was restricted because doing so would put them into competition with its customers. ALCOA favored being a semi-finished manufacturer selling to finished goods producers. Early on they had developed products and allowed or even created independent companies to market the finished goods such as foil, wire, and castings. Reynolds and Kaiser had to develop finishing operations to enter the market. By the end of the 1950s, ALCOA was forced to move into end product and finished goods production. This new strategic change required a new management approach.

ALCOA was still the dominant company after the war. However, most of the founding members were gone. Roy Hunt stepped down in 1950 as president, and I.W. Wilson became president. Roy Hunt moved on to play an active role as a board member. CEO A.V. Davis was in semiretirement and in his seventies. Andrew Mellon and his brother Richard were dead, but the Mellon family still had over 20 percent of the stock. The Mellon family was no longer viewed as the enemy of the operators. Mellon and Wilson would form an alliance that would mean new growth for ALCOA. I.W. Wilson would now have the opportunity to run the company, and he would be much different from the founders.

Since its founding, ALCOA had been a strange mix of autocratic rule and a decentralized structure. Captain Hunt, Charles Hall, Roy Hunt, A.V. Davis, and research head Francis Frary all had believed in this type of organization without an organizational chart. In fact, ALCOA had no formal organization until World War II, when the government required one for contract work. The Mellons, Andrew and Richard, played an active role in operating and expansion decisions, creating stockholder fights. There was strong interplay among shareholders and board politics. These individuals had strong personalities that often clashed.

"Chief" Wilson (1884–1954) had no direct ties to the company's founders and was not a major stockholder, but he was a product of ALCOA. Wilson graduated from MIT in 1911 and went to work in ALCOA's department of plant control and engineering at the Niagara, New York, plant. A year later, Mr. Wilson was transferred to New Kensington. It was here that he picked up the nickname "Chief." The name came from a popular Pittsburgh Pirates baseball player of the time. By 1921, he was general superintendent of ALCOA's primary reduction plants, which was the key operations job in the company. In 1931, he became the company's youngest officer when he was elected vice-president in charge of operations. He was elected a director in 1939, senior vice-president in 1949, and president in 1951. In 1956, he became chairman of the board and held that position to 1960. Wilson, like the founders, was an engineer and had a strong operations background. Wilson, however, was open to new alliances with the stockholders. He developed the old ties with the Mellon family, and particularly Richard King Mellon.

Richard King Mellon (1899–1970), commonly known as R.K., was in charge of the Mellon family stock. R.K. Mellon was the son of Richard B. Mellon, nephew of Andrew W. Mellon, and grandson of Thomas Mellon. R.K. Mellon had strong leadership skills, having served in the United States Army in both World Wars, receiving the Distinguished Service Medal, and rising to the rank of lieutenant general. Prior to World War I, he had entered Princeton; after World War I, Mellon returned to Princeton but left the university soon thereafter. He disliked formal education and hired tutors in subjects that would benefit him. Mellon set out to learn banking and finance from his father. When his father, Richard B. Mellon, died in December 1933, Richard King Mellon became president of Mellon National Bank and the Mellon representative on the ALCOA board.

After the death of his uncle Andrew in 1937, Richard King became the leader of the Mellon dynasty and board member of ALCOA. He also served on the boards of General Motors, Gulf Oil, and other Fortune 500 corporations as head of the Mellon fortune. Richard King Mellon proved to be the financer ALCOA would need. With many philanthropic interests, he let Wilson run ALCOA and completely backed him. R.K. served as a trustee of the Mellon Institute, the University of Pittsburgh, the Carnegie Institute of Technology (now Carnegie Mellon University), and the Carnegie Museum. As a member of the Pittsburgh redevelopment committee, he oversaw Pittsburgh's renaissance, which included tearing down almost 100-year-old buildings and replacing them with skyscrapers and modern office buildings. The ALCOA aluminum skyscraper would be one of those. As an ALCOA board member, he played a much more supportive role than his father or uncle.

The main change, however, was the evolution of a real corporate structure. ALCOA's evolution from a decentralized to a centralized corporation was long and painful. Middle managers had a type of loose and free structure. The structure would become more centralized in the 1950s. ALCOA had followed the path of Standard Oil and United States Steel in this centralized evolution. The first phase and intermediate step of this growth was the creation of committees to tie the decentralized functions to strategic goals. This system had mushroomed in the 1930s with many subcommittees reporting to corporate-level committees—the General Committee and the Technical Committee. The vision was to keep the "small company" feel. By the middle of the 1950s almost 100 committees were in place. ALCOA believed this type of system better tied together sales, research, and product lines. Chief Wilson would bring the company more in line with its large corporation identity. Eventually, these committees would form departments and centralize functions.

Wilson created new departments with executive-level managers. This included a Legal Department, a Design Department, a Public Relations Department, and an Industrial Relations Department. The Research Department was reorganized as well. Francis Frary retired in 1952; like his predecessor and founder, Charles Hall, had been very research-oriented on a personal level. Frary and Hall had allowed each plant to run its own research laboratory with little coordination. Wilson selected Kent Van Horn as director of research for his managerial skills. The plan was to centralize and coordinate research to support corporate goals. The structure was part of the evolution during the war, when engineers were more focused on current needs and problems versus creative long-term scientific projects so loved by the company founders. The change was reflected by a major increase in the research budget from $2.2 million in 1945 to $6 million in 1955.

Not surprisingly, Canadian spinoff Alcan had similar organizational problems. Interestingly, Alcan pursued a much different path in restructuring after the war. Alcan's structure had deep roots in the decentralized approach of its old parent company, ALCOA. The organizational change started at the end of the war before ALCOA stockholders had to fully divest. The war and the pending court consent to spin Alcan off completely from ALCOA left Alcan a much different company. First of all, it was truly an international company with major plants in Europe. Edward Davis came up with a new and unique style of international organization. Davis let his operating plants remain decentralized with executives of each operation evaluated on their performance. He then set up five group service companies: Aluminium Securities Limited, Aluminium Secretariat Limited (legal functions), Aluminium Fiduciaries Limited (public and employee relations), Aluminium Laboratories, and Aluminium Union Limited (sales). For a few years there was an operating company as well, but this was replaced with a director of operations, with operating heads reporting to him. The group companies were independent companies, who charged fees to the operating companies. Edward Davis believed this structure made these service companies earn their fees based on good service.

Edward Davis was focused on Alcan's status as a major international company. Davis realized that international managers with group companies would require a much different skill set. To this end, he established a school known as the International Management Institute in Geneva, Switzerland, to train his up-and-coming managers. A manager would have a year of studies there followed by multidiscipline and cross-cultural assignments. In 1947, Edward Davis decided to retire before the complete break with ALCOA stockholders. After 1947, his brother Arthur

Wedding of Richard King Mellon to Constance McCaulley, 1936 (Library of Congress).

Davis and the Mellons controlled the company stock; Edward was able to make his son Nathanael Davis president, though he was only 32 and had little company experience. While Nathanael was obviously an inside choice, he did have an excellent background to lead Alcan and its new organization. Clearly, Nathanael was groomed for the job of president. His father had sent him to Europe to study for a year before entering Harvard. After graduating in 1938 and spending some months with Alcan, he returned to Europe for a year of study at the London School of Economics. He would spend three years as a U.S. Naval officer in the Caribbean and Pacific war zones. In 1946 he resumed his career with Alcan in the London office.

The postwar international market and competition required all aluminum producers to look at their organizations. The Big Three of the United States, ALCOA, Reynolds, and Kaiser, made another change in organization to fit their market. The surplus of aluminum and fabrication plants after the war required a focus on new products and product development for Reynolds and

Kaiser. At the turn of the century, ALCOA had employed a unique approach of research and development. The approach was to develop new products and pioneer their production. After establishing the market, ALCOA would then pull out. It worked well when ALCOA had a market monopoly. After the war and the new competition of Reynolds and Kaiser, companies needed to break into ALCOA's fabricating customer base. All the companies looked to work with designers, fabricators, and end manufacturers cooperatively. Things had clearly changed; the product manufacturers and designers had become the customers for the aluminum companies. The aluminum industry had matured, and the various corporate roles resembled those of the steel industry. The key was to vertically integrate to manufacture semi-finished products such as wire, extrusions, tubing, and sheet. Investment needed to be focused on both increasing the efficiency of the vertical production and in the development of new products. Yet new products were needed to increase the demand for aluminum, creating tension between process and product developers.

To meet these two separate goals, the aluminum companies created a new design function and a new type of organization to support it. All three American companies developed design departments in the early 1950s. These design departments reported to the vice-president of sales or a planning committee. The old research and development function remained, but was more in support of operations. These new design departments, with their "clearly defined role for design in industrial planning,"[2] were a break from what used to be an R&D function and required a different type of employee. Still, the design department was also a link to research and development. The design department was actually a point person for outside designers. An outside designer could get information and help on alloys and their properties. If the designer needed something tested, the design department could make it happen.

The employees were characterized as follows: "The personnel in these design units came, for the most part, from a sales background or from design staffs of other corporate and industrial design offices, in contrast to the science and engineering backgrounds of most of the R&D staffs."[3] These design department employees were to work with designers in other industries to sell the advantages of aluminum. They were not to be designers, but to help designers. This was not new to the aluminum industry only, but to American industry. Still, aluminum had always been a product-oriented commodity. Products had, from its earliest history, defined the metal. The symbiotic relationship of art and industry was difficult to fully understand; but in the 1930s, the relationship had surfaced in the Art Deco movement. This symbiotic relationship would morph into the industrial design movement of the 1950s.

CHAPTER 15

The 1950s—The Products of the 1930s Come of Age

THE HISTORY OF ALUMINUM PRODUCTS is best studied from a design approach rather than a chronological one. For this reason, we must return to the 1930s and the Art Deco movement. While the price of aluminum had decreased dramatically by the 1930s, the metal was still very fashionable with artists and consumers. The Paris Exposition of 1937 was the world's celebration of aluminum. The Paris World's Fair and celebration would be an appropriate location. Paris was the city of aluminum design. The Paris Exposition of 1855 had introduced aluminum to the world, and the Paris Exposition of 1878 had introduced the key to its commercialization with Edison's electrical systems. The official name of the Paris 1937 Exposition was International Exposition dedicated to Art and Technology in Modern Life. The French aluminum association, L'Aluminum Française, sponsored a pavilion dedicated to the progress in aluminum. The exposition medal was made of aluminum. Politically, the German pavilion, designed by Albert Speer, was used to announce that Germany had overtaken America in aluminum production. Aluminum had overtaken steel as a symbol of power and technology. Unfortunately, many of these aluminum products exhibited would be put on hold during the war of the 1940s, but many of the Exposition's products would be commercialized in the 1950s.

The 1950s saw a new push towards aluminum as commercial jets turned to aluminum construction. Aluminum returned to its place in art and culture. It was the restoration of the aluminum age that had been interrupted by the war of the 1940s. In 1955, the 400-pound iron ball in Times Square was replaced with an aluminum ball weighing a mere 150 pounds. The 1950s saw an increase of aluminum in the construction of houses. The aluminum ice tray, Christmas tree, tumblers, and folding chairs changed the culture of suburban life. Kaiser Motors produced the all-aluminum concept car that many had dreamed about in the 1930s. The industrial design movement of the 1950s brought the merger of design, art, and industry.

Aluminum's use in design that emerged in the late 1930s was renewed in the 1950s, and the product history of the 1950s was deeply rooted in the 1930s. To understand the emergence of aluminum products in the 1950s, we go back to the future. Aluminum had caught the eye of many designers in the 1930s. Warren McArthur (1885–1961), whose furniture helped define the glamour of 1930s Art Deco curves, took aluminum furniture to high design. McArthur proved to be equal parts designer, artist, and manufacturer. In 1929 McArthur went out to Los Angeles and started a metal furniture business. He began making custom pieces but soon began to focus more specifically on creating new methods of joining aluminum pieces together. Aluminum, unlike steel, could not be welded properly in the 1930s. He developed notched tubes, washers, and screws to connect and strengthen aluminum tube furniture. McArthur also designed a unique joining system using circular rings, which made for a smooth appearance.

McArthur pioneered anodizing that would make aluminum hard and relatively impossible to tarnish. This anodized surface resisted scratches and dents, which allowed him to give his products a lifetime guarantee and to introduce colors. The anodic process made the metal porous, and the dyes infiltrated the pores in such a way that they actually became part of the metal rather than just an outer coat. The color tones that resulted, marketed in shades like Golf Green and Alice Blue, were incredibly popular and soon became an icon of 1930s Hollywood. Warner Brothers furnished their new theater in McArthur aluminum furniture, as did the Ambassador Hotel. McArthur's pieces were seen on the sets of movies and in the homes of stars and directors. They were recognizable for their curved tubing and their smooth, rounded edges. Unfortunately his company could not survive the economic downturn of the 1930s.

McArthur started a new operation in New York in the late 1930s. He used his engineering skills, and new joining methods allowed him to move into mass-produced cheaper chairs. During World War II, his company designed 500 models for aircraft seating and built an estimated 70 percent of all the aircraft seating for the Navy and Army Air Corps. The prolific McArthur authored 1,600 design patents. In 1939, McArthur designed what may have been the first aluminum folding chair. It was a true furniture chair versus the folding chairs of today. Just as important were the smooth curves and new colored surfaces that would reappear in the industrial design movement of the 1950s.

The 1930s proved to be a foundational movement of aluminum into building, art, and other applications. The economy hampered the development of many of these new applications that would resurface in the boom times of the

First aluminum building, 1931, Richmond, Virginia (Library of Congress).

1950s. In the 1930s, the New Kensington research operation developed a pewter-like aluminum alloy that started the Art Deco aluminum movement in tableware and dinnerware. New Kensington functioned as a type of skunk works for creativity and innovation in aluminum design of the 1930s. Many of the dreams of the 1930s would become reality in the 1950s. The visionary leader of this 1930s aluminum movement would be Buckminster Fuller (1895–1983). Fuller's ideas had to wait for the optimism of the 1950s. The war had brought a series of new alloys, which opened up new markets for old applications. Post–World War II excess aluminum capacity would spark a rush to new products as well. Unlike the 1930s, however, instead of artists, industrial designers would lead the new era of innovation.

The Big Three—ALCOA, Reynolds, and Kaiser—focused on design by organizationally adding design departments. The strategy changed from the aluminum companies' designing of new products to a cooperative effort with independent designers in the 1950s. Maybe just as important were the many innovations in areas such as aluminum housing and the transportation industry that had been blocked by the economic conditions of the 1930s. It wasn't just economics that slowed progress in the 1930s; it was the lack of new aluminum alloys capable of meeting product requirements. The war brought an array of stronger alloys that opened up new areas of products.

After the war, alloy 6061 was the structural workhorse of the industry. ALCOA looked to expand structural applications and products. Beyond aircraft and buildings, ships offered some opportunities. The softer alloys of the 1930s had often failed in shipbuilding applications. Alloy 6061 now had the strength. ALCOA highlighted its use in building ships for its Caribbean alumina mines. In 1946, ALCOA had built three ships to highlight the use of 6061 alloy. The steel hulls were those of the famous Liberty Ship freighters of World War II; otherwise, aluminum was used abundantly. In a speech to Society of Naval Architects, the symbolism of the ships was noted: "These liners, the S.S. *ALCOA Cavalier, Clipper, and Corsair* used alloy 6061 for deckhouses, bridges, smokestack enclosures, lifeboats, davits, accommodation ladders, hatch covers, awnings, weather dodgers, storm railings, all connected with 6065 rivets. Other alloys, both wrought and cast, were used for doors, windows, furniture, electrical fittings, décor and miscellaneous applications."[1] Shipbuilding and mining would lead to new ancillary business opportunities for aluminum companies.

Mining operations and its shipping line offered a unique opportunity for ALCOA in 1948. Mining cargo ships to and from the Caribbean could also act as travel ships for tourists. The SS *ALCOA Puritan* provided freight and passenger service between U.S. and Caribbean ports and a fleet of multipurpose ships was developed. An ALCOA ship was typically staffed with 10 officers and 33 crew members and could also accommodate 8 to 10 passengers in 1946. In 1948, most ships were upgraded, redesigned, and finished with accommodations for 96 first-class passengers. ALCOA hoped that by entering the passenger business, this would give the company a competitive edge on any rival shipping lines who might want to lure away some ore shipments.

These cruises were luxury level with air-conditioned suites, swimming pools, fine foods, dancing, and live entertainment. Stairs, rooms, and amenities used aluminum liberally. The cruise took 16 days from New Orleans to South America. The ship stopped at a number of South American ports to accommodate passengers. Some of the 10 passenger (less luxurious) ships went from New York and Montreal to South America. The ships were sailing advertisements for the use of aluminum. The author of *Aluminum Dreams* saw a cultural importance to these aluminum decorated and ore laden ships: "Thus the ships combined industrial mobilization of commodities, multisensory touristic mobile gaze, and subtle symbolic mobilization of the signs and icons of mobile modernity."[2] While ALCOA invested a lot in advertising and sales efforts, the cruises were ended in 1955.

The 1930s earlier had offered a similar merger of aluminum and culture. Earlier versions of Duralumin in the 1930s were used in bicycle frames. Aluminum bicycle frames went back to the 1890s, when they were made from cast aluminum. At the time aluminum alloys lacked the strength for tube frames, so thick castings were needed for strength. The development of Duralumin allowed for tube frames. Antonin Magne (aka "The Monk") won the Tour de France in 1931 and 1934, when he secretly tested the first Duralumin rim during the race. Aluminum rims became the standard for road racing by 1937. Magne even went so far as to paint wood grain on the rims to avoid detection. A number of bicycle companies soon produced aluminum bikes with mechanically connected tubes. They commanded a high price compared to welded steel tube frames. Aluminum welded tube framing was not possible until the 1950s when better welding techniques were developed for these stronger alloys. Still, aluminum offered better corrosion resistance than steel and significant reduction in weight over steel. The aluminum alloy Duralumin offered equivalent strength and dent resistance to steel.

The late 1930s brought a new interest in airships because of this improved aluminum tubing. Actually, dirigibles made by Count Zeppelin had continued their development. In 1919, Germany launched the first passenger airship, the *LZ-120 Bodensee*. The German Zeppelins were the perfect embodiment of art Deco in the air. The airship had been a feared military weapon in World War I. Things changed with the airship *Bodensee*. While Zeppelin's early flights were sightseeing tours, the *Bodensee* began scheduled service between Berlin and southern Germany in 1919. The flight from Berlin to Friedrichshafen took 4 to 9 hours, compared to 18 to 24 hours by rail. *Bodensee* also carried cargo and the mail. In 1931, the Germans re-established the world's first passenger airline, DELAG (Deutsche Luftschiffahrts-Aktiengesellschaft, or German Airship Transportation Corporation Ltd).

The German airships were a blend of iconic 1930s design and the use of aluminum. The *Hindenburg* was the top of the line. The *Hindenburg* had a Duralumin structure, incorporating 15 Ferris wheel–like bulkheads along its length. Sixteen cotton gas bags were fitted between these, and the Duralumin bulkheads were braced to each other by longitudinal girders placed around their circumferences. The airship's skin was of cotton doped with iron oxide and cellulose acetate butyrate impregnated with aluminum powder. The aluminum powder was added to both reflect ultraviolet, which damaged the fabric, and infrared light, which caused heating of the gas.

The *Hindenburg* made 17 round trips across the Atlantic Ocean in 1936 with 10 trips to the U.S. and 7 to Brazil. In July 1936, the airship also completed a record Atlantic double crossing in 5 days, 19 hours, and 51 minutes. After defeating Joe Louis, the German boxer Max Schmeling returned home on the *Hindenburg* to a hero's welcome in Frankfurt.

German designer Professor Fritz August Breuhaus used lightweight tubular forms to create a public space where people were expected to spend their time instead of in cramped cabins. The German airships were known for their aluminum furniture. The tables and chairs were designed by Breuhaus using lightweight tubular aluminum, with the chairs upholstered in red.

The *Hindenburg* was the greatest of the class, originally built with 25 double-berthed cabins accommodating 50 passengers. The *Hindenburg* featured the first aluminum (Duralumin) piano ever to be carried on a passenger aircraft. It weighed 360 pounds and was covered with yellow pigskin. The *Hindenburg* also started transatlantic service to South America. Zeppelins were faster than ocean liners. The Empire State Building was completed in 1931 with a dirigible mast, in anticipation of passenger airship service. Germany's success led to the term *zeppelin* being genericized from the original trademark that referred to airships manufactured by the German Zeppelin Company. Airship production in Germany would become the source of future stronger alloys and structural extrusions used in the 1950s.

The American experience with airships was much different in the 1920s and 1930s. The United States had only manufactured three rigid airships in the 1920s. These airships were built by Goodyear, and all three met with disaster. The USS *Shenandoah* flew into a severe thunderstorm over Noble County, Ohio, while on a poorly planned publicity flight in 1925. It broke into pieces, killing 14 of its crew. The USS *Akron* was caught in a severe storm and flew into the surface of the sea off the shore of New Jersey in 1933. It carried no lifeboats and few life vests, and its crew of 76 died from drowning or hypothermia. The USS *Macon* was lost after suffering a structural failure offshore near Point Sur Lighthouse in 1935, but only two crew members died, thanks to having life vests. Still, these accidents ended the airship in America. Germany would follow suit after the crash of the *Hindenburg* in 1937, which killed 35. Had the *Hindenburg* been filled with helium, as opposed to hydrogen, the disaster would not have occurred. The aluminum airship alloy tubing and sheet of the 1930s nonetheless set the future for many products of the 1950s.

Architecture became another opportunity for aluminum in the 1950s. The iconic all-aluminum skyscrapers of the 1950s were deeply rooted in the aluminum architecture of the 1930s. Art Deco of the 1930s returned as the industrial design of the 1950s. Art Deco emerged in the 1930s when rapid industrialization was transforming culture. The style takes its name from the Exposition Internationale des Arts Decoratifs held in Paris in 1925 as a showcase for new inspiration. Its roots were also in the artwork of Jules Verne books, which added rich Victorian decoration to machinery. One of its major attributes of Deco was an embrace of technology. This distinguished Deco from the organic motifs favored by its predecessor Art Nouveau. It was an eclectic style that combined traditional craft motifs with Machine Age imagery and materials such as aluminum.

In 1930, aluminum and Art Deco merged in the Times Square Building (the former Genesee Valley Trust Building). The Times Square Building is a skyscraper located in Rochester, New York, designed by Ralph Thomas Walker; at 260 feet, it is the 8th tallest building in Rochester, with 14 floors. The building is a streamlined 12-story structure supporting 4 aluminum wings 42 feet high, known as the "Wings of Progress," each weighing 12,000 pounds. These structures are among the most distinctive features of the Rochester skyline. The first extensive use of aluminum in construction was the Art Deco Empire State Building in 1931: the entire tower portion was cast aluminum, as well as many decorative features such as the entrances, elevator doors, ornamental trim, and some 6,000 window spandrels. Since strong alloys were lacking in the 1930s, thick castings were required.

The cast aluminum tower on the Empire State Building used a new alloy developed at ALCOA's Cleveland plant. The new cast alloy would not only be a mainstay in Art Deco architecture but in industrial products. This alloy, known as A195, was used in aluminum cast pistons and aircraft engine blocks in the early 1920s. A195 could be heat-treated to improve strength. In 1928, the 11,340 tons of heat-treated cast products also included washing machine agitators, vacuum cleaner bodies, and food processing equipment.[3] Many of these industrial and consumer cast aluminum products were designed in Art Deco and are credited with the creation of the industrial design movement. In Chicago, the nightime skyline features the aluminum alloy sculpture of Ceres by John Storrs that tops the central tower of the Chicago Board of Trade Building (1930). This type of cast alloy would become commonplace in government buildings of the 1930s.

In 1931, an innovative but short-lived use of aluminum as a building material was made at the suggestion of James Bolton, a city official in Richmond, Virginia. An aluminum rectangular building was finished in late 1931. Locals hailed the structure as the nation's first all-aluminum building. Materials were provided by ALCOA. It was designed to offer temporary and inexpensive space for the city's Department of Public Works, but it had the looks of a giant three-story trailer. Its boxy appearance fit no specific architectural style. It was looked at as a temporary government office box lacking any architectural value. The local government was able to relocate the entire building to an out-of-sight location in 1939. The concept never took off. Office buildings soon turned to the permanency of Art Deco.

Not surprisingly, another example of Art Deco and aluminum in architecture was the Department of Commerce Building in Washington, D.C., officially known as the Herbert C. Hoover Building. Designed by Louis Ayres and constructed between 1927 and 1932, the Commerce Department building is named after President Hoover who served as Secretary of Commerce during the agency's early development. Of course, his Secretary of Treasury was Andrew Mellon. It was the largest office building in the world at the time of its completion in 1932. It was a Classical Revival–style building, but distinguished by Art Deco architectural elements and its innovative use of aluminum for details. All entrances to the building feature 20-foot-high aluminum doors that slide into recessed pockets. Interior stair railings, grilles, and door trims are aluminum, as are Art Deco torchières, doors for the building's 25 elevators, and more than 10,000 light fixtures.

Examples of Art Deco architecture and aluminum can be found in most American cities. The First Center for the Visual Arts (former post office) in Nashville is another beautiful example. Constructed in 1934 by the local architectural firm of Marr & Holman, the building was financed by monies appropriated by Congress during the Hoover administration. Following guidelines from the Office of the Supervising Architect for such public buildings, Nashville's post office displays the most distinctive architectural styles of the period, such as cast aluminum doors and grillwork, as well as colored marble and stones on the floors and walls. The Durham Museum in Omaha, Nebraska, shows similar beauty in its aluminum Art Deco, including the addition of aluminum leaf that once was the wall covering of the very rich in the 1890s. In Seattle, spectacular aluminum screens and doors comprise the entry at the Seattle Asian Art Museum. Other examples include Cincinnati's old Union Terminal Building, plus the Toledo Library; almost every major American city has examples. These buildings show the visual beauty of aluminum Art Deco. This beauty led to other applications.

ALCOA moved directly into Art Deco with tableware and giftware known as Kensington Ware in 1934. Kensington Ware, known as "the poor man's silver," was used as fancy giftware given at weddings and other special events, and was often displayed prominently in dining rooms. Kensington Ware was a heavier pewter-like alloy of aluminum. Like modern pewter, Kensington maintained its finish. ALCOA's Kensington Ware first came to the public eye in the summer of 1934, with showings at the Chicago and New York Gift Shows, attended by over 4,000 retail buyers. A Kensington sales brochure from early 1935 touted the line's numerous advantages as having most of the advantages of silver, pewter, and chromium, yet none of their disadvantages. It will not tarnish or stain, but retains its soft, silvery luster. It also had a heavier feel than most aluminum alloys. Kensington Ware was to be of heirloom quality, yet priced below silver at $10 to $20 in the 1930s. Some items such as picture frames, cigarette boxes, and candlesticks were under $10. Today Kensington Ware is highly collectible with prices from $80 to $200. Pieces by famous designer Lurelle Guild can sell much higher today.

Lurelle Guild, whose corporate clients included Westinghouse and Electrolux, became a designer for ALCOA. He developed a relationship with ALCOA in the 1930s, helping to modernize the "new" material of aluminum through his designs for a stylish museum and showroom in New York and a line of products. He often created objects of old brass and made Kensington Ware popular with artists. Guild produced his own line of ALCOA's Kensington Ware. Lurelle Guild's streamlined restyling of the Electrolux vacuum cleaner in 1937 using Kensington metal would make him one of America's top industrial designers. Guild and ALCOA maintained a Kensington Ware permanent showroom in New York's Rockefeller Center.

Lurelle Guild was not alone in seeing the potential of high-quality aluminum giftware. In 1930, a commission to create decorative aluminum gates and elevator doors for its research center in New Kensington led to a new era of design innovation of forged aluminum projects. In nearby Grove City, Pennsylvania, Wendell August Forge had been working in forged art pieces and was contacted to make the gates for ALCOA. Wendell August had hired Ottone Pisoni, former blacksmith, to develop aluminum forgings (also known as hammered aluminum) for artwork. It was a true skill to be learned. Pisoni mastered the art of forging aluminum, and the company prospered. Employees from Wendell August went on to open competing businesses by the mid–1930s.

The Kensington line of tableware would make a comeback in the 1950s. However, aluminum products were produced in bright colors to fit the culture of the 1950s. Giftware such as the forged aluminum products of Wendell August grew in popularity with new pewter-type alloys. The 1950s saw a reuniting of culture and aluminum in giftware. This new unification of culture and aluminum was more utilitarian, moving away from pure art to design. But it was really the culture of the 1930s impacting aluminum products.

The cultural impact of aluminum had also been seen throughout the 1930s. Industry and engineers became aluminum's biggest supporters. The 1939 World's Fair featured some of the last consumer aluminum applications prior to the war. Elektro was the nickname of a robot built by the Westinghouse Electric Corporation in its Mansfield, Ohio, facility between 1937 and 1939. It was on exhibit at the 1939 New York World's Fair with "Sparko," a robot dog.

Elektro was a seven-foot-tall, brushed gold-colored aluminum robot that weighed 265 pounds. Humanoid in appearance, it could walk by voice command, speak about 700 words, answer questions (with behind-the-curtain help), smoke cigarettes, blow up balloons, and move its head and arms. Elektro became the most famous robot of the 1930s and a model for movie robots of later decades. Elektro starred in the 1939 film, *The Middleton Family at the New York World's Fair*. After the war, Elektro toured North America in 1950 in promotional appearances. Ten years later, Elektro (under the stage name Thinko) starred with Mamie Van Doren and Tuesday Weld in *Sex Kittens Go to College* (also known as *Beauty and the Robot*), a 1960 American comedy film.

Probably the most unusual blending of art, culture, politics, and aluminum that occurred in the 1930s was in Italy. Fascist Italy was in search of an identity in the prewar Europe of the 1930s. Particularly, Mussolini was interested in creating a nation of modernity and technology. Italy, however, lacked the raw materials of the earlier Victorian era of coal, iron ore, and petroleum. Italy had large resources of bauxite in Istria, Campania, and Sicily. Mussolini and the fascists seized on the idea that aluminum would be the symbol of Italy's rising strength. His brother's famous quote was, "Just like the 19th century was the century of iron, heavy metals, and carbon, so the twentieth should be the light metals, electricity, and petroleum.... If we [Italians] haven't iron, we have aluminum."[4] The fascist literature stated that aluminum "embodies Italy's unyielding destiny." Mussolini had aluminum statues made to be exhibited in Rome's Olympic Stadium and Italy's pavilion at the Paris International Expo of 1937. While Mussolini created an aluminum art movement, the country never achieved its goal of a major aluminum industry. By the end of the 1930s, Italy was far behind Germany and France, and even ranked below Hungary and Yugoslavia in aluminum production.

One lasting monument to Italy's aluminum dream and the art of the 1930s is the Moka Express coffeemaker of designer Alfonso Bialetti. Bialetti had worked for years in the French aluminum industry and returned to Italy to open a metal shop. Bialetti started to use cast iron molds to cast aluminum. The cast iron molds were reusable and cheaper than sand casting. In 1933, he merged his aluminum craftsmanship with his inventive genius to produce a home espresso coffee maker. The aluminum design and the popularity of espresso coffee made for good sales; however, Bialetti's small shop limited production. At best he was able to produce 10,000 a year. By 1940, the war ended the use of aluminum for consumer products. After the war in 1954, Bialetti moved to a state-of-the-art factory, and production and sales boomed. The espresso machine's sales increase came even with the high cost of Italian aluminum. Today, the factory does well, selling over three million a year of these aluminum coffee machines. These Moka Express machines are produced by the same methods used in 1933. Bialetti proved the importance of craftsmanship, design, and metalwork.

The Italian experience of aluminum blending into culture and national identity was not new nor has it disappeared. Germany and France had, in the early 20th century, adopted aluminum as a national representation of power and technical progress. This would be followed by Russia in the 1990s, China in the 2000s, and most recently by Australia. Australia's position today as the world's largest bauxite producer has been reflected in their very national identity, as seen in an art movement within the country that reflects its prominence in the global aluminum industry.[5] Australia leads in new domestic applications of aluminum as well as promoting aluminum art today.

Clearly, aluminum in the 1930s had a deep effect on culture, art, and industrial design. It was the golden age of aluminum design. The aluminum car tried again to make a comeback in 1933, using sheet metal versus the heavy cast bodies of previous decades. The Dymaxion car was a three-wheel concept car designed by U.S. inventor and architect Buckminster Fuller in 1933. The word Dymaxion is a brand name that Fuller gave to several of his inventions, to emphasize that he considered them part of a more general project and to improve humanity's living conditions. The name comes from **DY**namic, **Max**imum, and tens**ION**. The Dymaxion car first caught the public's attention when he lent it to his friend, Amelia Earhart, for a visit to the White House. The car was shown at the 1933 Chicago World's Fair. Actually it was more of an SUV than a car. It could seat 11 people, had a recorded top speed of 90 miles per hour, with a rear-mounted, 85-horsepower Ford flathead V-8 engine, which sat near a single rear wheel and powered the car's two front wheels. It was the first application of Art Deco to industry; in fact, Fuller got the idea from architecture. It was a highly advanced, aerodynamic car with slim lines. The car had a fuel efficiency of 30 miles per U.S. gallon, which was amazing for the time. Fuller's car lost some credibility with the public when it was involved in a fatal crash.[6] Ford Trimotor airplane designer Harry Stout tried creating his own version of the Dymaxion aluminum car in 1934. He used an Art Deco design also to streamline the body and increase mileage. Like the Dymaxion, it was more of a minibus or minivan. The 1930s proved a poor time for such innovative designs. In 1937, BMW produced a beautiful aerodynamic Art Deco aluminum racing car, but the approaching war demands in Germany ended production. Mercedes-Benz Classic 540K Streamliner in 1938 also had a beautiful aluminum body. Only one version was made with aluminum-alloy bodywork, and these were made for high Nazi officials. Like America, the war ended Germany's expansion into aluminum cars.

New Kensington plant, 1938 (Library of Congress).

After World War II and the Korean War, there was a surplus of aluminum in Europe. This surplus was created by overproduction during the war, and the slowness of consumer markets to bounce back in war-torn Europe. Stronger aluminum sheet was now available. The Panhard Dyna Z was an aluminum car produced by Panhard of France from 1954. It had aluminum body skin but the frame was steel. The decision to use aluminum sheeting had been made at a time when a sudden dropoff in demand for French and Allied fighter planes had left the producers with a glut of the metal in the early 1950s. By the time the design went into production, the relative cost advantage of sheet steel had increased steadily. By late 1954, the cost penalty of progressing with aluminum bodywork had become financially prohibitive and the design was switched to steel. It seems that in every decade since the 1920s, an effort has been made to resurrect the aluminum car. Price versus technology had now become the bigger problem in the 1950s.

By the end of the 1930s, aluminum was making inroads in land travel in other ways. Greyhound Bus in 1935 introduced a fleet of aluminum buses built by General Motors. The Art Deco streamlining was used to produce aluminum trains by the end of 1939. France in 1935 launched the ocean liner, which had 12 all-aluminum luxury cabins. Many of the furnishings were aluminum as well.

Aluminum seats were being put in railroad cars. These applications expanded with the stronger alloys of the 1950s.

Another application was the aluminum streetcar, which was the core of urban transportation in the 1930s. The Pullman-Standard all-aluminum car was completed at the Chicago plant in 1934 in Art Deco style. Its most outstanding feature was that it was built entirely of cast aluminum. A 50-foot streetcar carrying 58 passengers, it weighed only 29,600 pounds—only 55 percent as much as the 44-seat Old Pullmans that held down service on many of Chicago's heavier lines in the 1930s. Other cities experimented with aluminum streetcars, but they were hard to make from sheet aluminum. Steel remained the economical choice. The cost of aluminum sheet fabricating remained a roadblock into the 1950s and beyond.

The first use of aluminum in the railroad market was by the New York, New Haven, and Hartford Railroad in 1894 when it built a special lightweight car with aluminum seat frames. The first aluminum hopper car was built in 1934 by the Baltimore & Ohio Railroad. In the 1930s, both the Union Pacific and the Santa Fe railroads introduced aluminum passenger cars. The 1933 Union Pacific *M-10000* was a revolutionary train. Made of the aluminum alloy Duralumin, the three cars together weighed only 85 tons, while a conventional 10-car steam train weighed roughly 1,000 tons. The *M-10000* was as much a publicity tool as a practical train. During 1934 it made a 13,000-mile cross-country exhibition tour, visiting Washington, D.C., for inspection by Franklin Delano Roosevelt. Everywhere it went it attracted crowds and press attention, hosting almost 1,000,000 visitors. In 1935, the *Comet* full aluminum train was commissioned by the New York, New Haven, and Hartford Railroad and was built by Goodyear Zeppelin Corporation. Goodyear was looking for a new use of its aluminum airship factory after headline-grabbing crashes had soured the public on airship dirigibles. The full use of aluminum railcars would wait till the end of World War II and a surplus aluminum market. Interestingly, the *M-10000* was scrapped for its aluminum in World War II.

While the war years of the 1940s had stopped the commercial development, the surplus of the postwar years would foster a creative boom. The late 1940s would see the evolution of unique aluminum applications. The use of this surplus would create a new industry of aluminum housing and construction products. At first it was temporary housing for the war-misplaced population of Europe. In America, it was cheap housing for the returning veterans and their new families. This was one of America's greatest housing booms in the 1940s. Housing applications offered a new market for new producers such as Reynolds and Kaiser.

The design of the Dymaxion car in the early 1930s had interrupted one

ALCOA Building, 1958 (Library of Congress).

of Bucky Fuller's biggest visionary projects. It was part of Fuller's one-world vision of an earth connected by airships with people traveling and moving often to new locations. The vision had evolved in the prosperity of the 1920s. Fuller had been studying the development of an aluminum house that could readily be moved. This had come out of his vision of "One Town World" connected by airports for aluminum dirigibles or small planes. These dirigibles or planes would be used for mid-range flights with stops in remote area ports. These stopover ports in very remote areas would be for passengers to rest. He envisioned self-sustaining, easily assembled houses that would run off solar and wind power.[7] Fuller hoped to avoid massive construction sites by airdropping in these houses. He also wanted to eliminate the need for maintenance and support people at these sites. The idea of self-sustaining mobile houses took him into the deep study of how people live and meet their basic necessities.

In 1929, Fuller formed the 4D Company to sell these autonomous and transportable aluminum houses. These were much different from today's mobile homes. First, they looked like a chocolate Hershey Kiss more than a traditional home. The 4D houses utilized passive solar and wind for heating and cooling. They also had fireplaces and separate rooms for the kitchen and living room. The 4D Houses were also meant to be produced on an assembly line and marketed as a home for the average American. For example, Fuller designed a mass-produced bathroom to replace the costly on-site construction using plumbers and fixtures. His study of toilets took him to manufacturing plants all over the country. His new aluminum and stainless toilet did get serious consideration from many manufacturers.

He based his overall design on automobile costs. His fully furnished, two-bedroom, two-thousand-square-foot family home would cost $1,500 (about $17,000 in today's dollars). This was at a time when the average unfurnished house cost $8,000. He based his $1,500 on a 6,000-pound weight at 25 cents a pound, which was roughly the automotive standard. In addition, Fuller suggested a moving company would deliver and assemble the house anywhere in the world within days.[8] Fuller was an excellent engineer and had detailed drawings and specifications.

One problem Fuller had in the late 1920s was that he was penniless and lacked creditability. He operated out of Rosemary Marie's tavern in Greenwich Village, giving talks about the project. He also made arrangements with Rosemary Marie's to redesign it for a payment of one meal a day. Not surprisingly, he used aluminum fixtures. At the time he was homeless and often lived with friends and even slept on the streets. Fuller believed this modern subsistence living helped him to design a lean house. He did catch the ears of professional architects, lecturing at Yale University Architectural School, Princeton School of Architecture, and Carnegie Institute of Technology. At times, Fuller's visionary plans stretched the imagination. For example, he envisioned dirigibles dropping bombs to crater out the foundations for his houses.

Fuller was not alone in his use of aluminum in houses. Architects A. Lawrence Kocher and Albert Frey were asked to design an aluminum-based home for the 1931 Architectural League Exhibition. They designed the Aluminaire

House as an affordable, easy-to-reproduce unit for mass production. Its name is the interplay of the aluminum building material and light. This was a permanent structure, prefabricated, not mobile as in Fuller's case. Three-inch-thick aluminum walls sealed three stories of space supported by six slender aluminum columns secured with screws and washers. Glass and corrugated aluminum sheathed the outside. Frey's corrugated siding would be popular in building and would lay the foundation for the future of aluminum siding in the 1960s. Another aluminum building visionary was French architect Jean Prouve. Prouve had practical experience in the building of prefabricated barracks-type buildings for war refugees. Ultimately, Frey and Prouve suffered the same roadblocks as Fuller.

Fuller's timing was off by at least a decade. The Great Depression of the 1930s would dry up investment capital for such far-reaching projects. Fuller had met with ALCOA engineers about the stronger alloys needed in 1929. He would be at least 8 years ahead of the stronger alloys such as 61S needed for his design. Fuller realized, by 1933, his house project would have to wait for later times. Fuller's house design would surface again in the 1940s in the Dymaxion House, which offered an outlet to war aluminum surpluses. The Dymaxion House had all the characteristics of his earlier 4D house but with a cost of $6,000. The Dymaxion was more rounded than previous designs. The Dymaxion's round shape minimized heat loss and the amount of materials needed. The Dymaxion never really took off, although the traditional housing market exploded with the GI Bill. There remained a strong market resistance to this type of housing.

Fuller's concept of mobility was too cutting-edge for the time. Home designer Foster Gunnison, who was involved in the construction of New York City's Radio City Music Hall and the Empire State Building, was interested in becoming the Henry Ford of housing. He planned to mass produce houses on an assembly line. These houses, or actually kits, would then be put on traditional foundations. In 1936, he opened Gunnison Housing Corporation in New Albany, Indiana. Gunnison designed an interchangeable aluminum-framed and wood-panel wall that would fit 12 different house models by 1937. Gunnison could undersell a conventionally constructed house by almost 25 percent. Sales were slow in the late 1930s, but homes were shipped throughout the country. The war put limitations on the amount of aluminum available, but in 1942, the Lanham Act provided funds for war housing and prefab housing for defense industry workers. A number of prefab companies, including Gunnison Homes and National Homes, gained government contracts worth $153 million to provide 70,000 homes. Steel became the cheaper cost material; and eventually, in the 1950s, United States Steel purchased the plant. The concept never was fully accepted by the American public. Housing proved to be one area where aluminum could not break traditional thinking and images.

Europe was much more open to aluminum housing, and prefabricated homes found demand in England after the war. England developed a government-supported housing program to promote prefab aluminum houses utilizing old aircraft factories. By 1949, over 75,000 homes were produced, but government funds ran short and the program was canceled. In addition, the Bristol Aeroplane Company developed a prefab construction system for schools and similar buildings.

One area of success in prefabricated buildings in the United States was the aluminum diner. The aluminum diners were originally modeled after rail car diners of the 1930s. These diners tended to have the streamlined Art Deco design of the 1930s. The war stopped the growth of diners, but they made a major comeback in the late 1940s and 1950s. While built to be mobile, the aluminum diner tended to be a semi, if not a fully, permanent building. The fast food industry would end the era of aluminum diners by the end of the 1950s. The 1950s would usher in a new array of products and applications related to the culture and the postwar boom.

CHAPTER 16

The 1950s—The Aluminum Age

THE 1930S AND 1950S WERE INTERRELATED in aluminum usage and products. The 1940s, of course, had given aluminum new strength. Still, the 1950s also left its own mark on the aluminum industry. Aluminum entered a utilitarian era, often leaving behind its early history as a precious metal and its strong connection with art. In the 1950s, aluminum became a structural metal as well as a mainstay in the middle-class home. Its pricing reflected that of a commodity versus a special metal. The strong war alloys opened new markets in construction. Artistic tableware gave way to mass production. It's not surprising that the decade would start with the construction of the iconic all-aluminum ALCOA headquarters.

Aluminum did make major inroads in heavy construction with the building of the ALCOA skyscraper in Pittsburgh in 1950. The Aluminum Company of America was considering a move to New York in 1949. In an effort to keep the headquarters based in Pittsburgh, the Mellon family offered to build a parking garage and plaza adjacent to the proposed building. The company remained in Pittsburgh, and Mellon Square Plaza was constructed in the city block between the ALCOA Building and the Mellon Bank Building. The ALCOA building is a thirty-story skyscraper using the best of aluminum technology. Like Fuller's dome, the ALCOA building was a revolutionary use of aluminum in architecture. The unique aluminum walls of the building are only ⅛-inch thick. The windows could be turned 360 degrees for easily cleaning. The entire façade of the tower is sheathed in stamped aluminum panels of a special aluminum-silicon alloy. The aluminum was anodized to a gray color. The windows, sashes and frames, heating and ventilating ducts, water piping and wiring system were all made of aluminum. The use of aluminum internally had only been tried before in Bucky Fuller's Dymaxion House. Architects hailed and noted: "The erection of the ALCOA Building provided incontrovertible evidence that what architects like Mies van der Rohe had dreamed about in the

1920s and 1930s—the industrialization of building—was now an accomplished fact."[1] Mies himself used aluminum in luxury homes and skyscrapers such as the IBM Building in Chicago in the 1970s.

Probably the biggest advance for aluminum in the architecture of the ALCOA building was the curtain wall. The first aluminum curtain was in Pietro Belluschi's Equitable Building in Portland, Oregon, in 1948. A curtain wall system is an outer covering of a building in which the outer walls are nonstructural, but merely keep the weather out and the occupants in, and have an aesthetic value. Curtain walls are typically designed with extruded aluminum members, although the first curtain walls were prepared of steel. The 1950s saw a boom in the aluminum curtain wall design and expanded the use of aluminum in all types of housing. Reynolds Aluminum took leadership in the promotion of aluminum in architecture by working closely with architectural schools. Reynolds offered rewards for the best in aluminum architectural design and supported research in aluminum engineering. The engineering and architectural outreach would change the nature of how aluminum companies did business.

Kaiser Aluminum and Bucky Fuller would team up in the 1950s to further expand the use of aluminum. Fuller had had another shot at the aluminum car in the 1940s. In 1942, Henry J. Kaiser entered the automotive field and commissioned Buckminster Fuller to design a car. That venture didn't work out, but the two collaborated again several years later to explore the commercial potential of geodesic structures as a project of Kaiser Aluminum. Fuller did find a commercial success in his famous aluminum geodesic dome in 1954, which the military quickly embraced. For once, Fuller's timing was perfect, as art moved out of the Art Deco design into modernism, and the military through the war had become part of the aluminum industry. Geodesic domes found many applications with the military. These light but extremely strong buildings could be airdropped into dangerous and rough terrain. The geodesic dome was also embraced by traditional architects. Fuller also received credibility with many successful non-aluminum inventions that allowed him to move confidently back into aluminum.

Henry J. Kaiser installed a geodesic dome at the entrance to his Hawaiian Village Hotel on Kalia Road in Waikiki. The Hawaiian Village Dome had 16,500 square feet of covered area. It was 45 feet in diameter and was 49½ feet high. It was built in a remarkable 20 hours after the crew predicted it would take 5 days. The dome's construction efficiency was predicated on the benefits of prefabrication—a process perfected by Kaiser and his workers in the World War II shipyards and the building of Liberty ships. The aluminum panels were fabricated at the Kaiser plant in Permanente, California, and shipped to the construction site. The unconventional modular assembly process worked smoothly beyond expectations. They finished before Henry Kaiser was able to fly out from California to see it being built. Despite missing the action, he was proud of the workers. The completed dome was christened with a showing of Michael Todd's *Around the World in Eighty Days* in early November 1957. The dome was demolished in 1999 to make way for the Kalia Tower, which opened in 2001. Following the success of these projects, Kaiser and Fuller secured contracts for more domes and even set up a dome sales office in Chicago in 1958.

In 1960 a geodesic aluminum dome was constructed at the headquarters of ASM International, formerly the American Society for Metals. The dome is the world's largest open-air geodesic dome. It was built by Kaiser Aluminum and was made of extruded aluminum pipe; the open-work dome stands 103 feet high and 250 feet in diameter, weighs 80 tons, and contains more than 65,000 parts. The alloy used was 7075, the same secret alloy first used in the Japanese Zero and today in Apple's iPhone. The dome stands on five pylons, two of which rise up from courtyards set into the building. The American Society of Metals chose an aluminum dome because it stood as a symbol of humanity's mastery of materials: from the minerals and laboratories they come from, to the processes that make them useful in ways that touch our lives every day. For metallurgists, aluminum, unlike steel, required a massive confluence of technological advances. Most of America would see the Fuller aluminum geodesic dome at the 1964 New York World's Fair.

The 1964 fair, like so many world fairs since 1857, highlighted the metal aluminum. Besides the aluminum geo dome, there was a spectacular Tower of Light. The Electric Power and Light Companies' pavilion featured a 12-billion-candlepower welcome light that could be seen for miles in every direction. It was brighter than 50 illuminated Yankee Stadiums or 340,000 automobile headlights. At night, the hundreds of aluminum panels were bathed in ever-changing colored lights; during the day, the building reflected the sun's rays with colorful iridescence. Even the fair's symbol, the Unisphere, was originally designed in aluminum, but at the last minute was changed to stainless steel. Still, in so many ways this was aluminum and Bucky Fuller's fair. It would bring about a renewal of aluminum in transportation in car trailers.

Bucky Fuller's commercial failures had become the future dreams of a new line of innovators. One mobile use of aluminum did survive the 1930s to become a popular product. In 1936 Wally Byam introduced the "Airstream Clipper," an aluminum trailer. The design cut down on wind resistance and thus improved fuel efficiency. It was the first of the now familiar sausage-shaped, bright aluminum Airstream trailers. Around 1935, Fuller became associated with William Hawley Bowlus, who was a pioneer in airplane design and builder of the historic airplane, the *Spirit of St. Louis*. At the time, Bowlus was building a trailer of stressed aluminum. The alloy was Duralumin, and Bowlus was aware of the problems of welding in the Junkers aircraft. Bowlus adapted the rivet construction of aircraft. He wanted Wally Byam for his sales and marketing expertise. Due to financial troubles, Bowlus declared bankruptcy in 1936. Byam purchased the company and improved on the design.

The first Airstream, called the "Clipper" in 1936, was named after the first transatlantic seaplane. It used principles of Art Deco design. The Clipper slept four, carried its own water supply, was fitted with electric lights, and cost $1,200. Airstream found popularity even in the Depression. During World War II, travel became a luxury and most could not afford trailers. The nation faced an acute aluminum shortage, and manufacturers restricted nonmilitary applications. When World War II ended, the economy boomed, and people's attention once again turned towards leisure travel. Airstream went back into production in 1948, and its popularity boomed in the 1950s. Even the aluminum car would resurface in a concept car of Kaiser Aluminum in 1958, and variations continue to this day, such as Ford's F150 truck.

Airstream's success in the 1930s did not translate into success of similar products in more permanent living structures. Harry Stout in 1936 invented a variation of the Dymaxion House. Stout had been the designer in the 1920s of the aluminum plane, the Ford Trimotor. Stout used the Dymaxion idea and played off Airstream. Like Dymaxion, however, it was meant to be semipermanent, not a trailer home. While consumers had rejected permanent aluminum housing, the success of Airstream had opened them up to semipermanent trailer homes. Stout also tried a fully mobile home in 1936, which many consider the first minivan. The fully mobile home never caught on, but it laid the groundwork for the future minivan. The Stout minivan utilized aluminum in the body and frame. The 1930s were clearly not the time for such innovations. Some of the aluminum features in housing such as siding would eventually catch on in the 1950s. Also, the product we call a mobile home did find a market in permanent housing, and it was based on aluminum siding.

In 1947, Jerome Kaufman invented a process to bake enamel coating onto aluminum siding and formed the Alside Company. Initially, colors were limited to white, gray, and cream, but additional colors were added. The siding came in smooth or textured surfaces, with or without an insulated backing. The building sector responded to these innovations, and by the late 1940s, aluminum siding was favored as an inexpensive material that could be installed easily and quickly. Reynolds Metals promoted the material as rot and fire resistant, offering it in panels, clapboards, and weatherboards. First producers of aluminum clapboard siding was Frank Hoess, a machinist from Hammond, Indiana. In

1946, Hoess joined forces with Metal Building Products of Detroit, a new corporation created to promote and sell Hoess-designed clapboard interlocking siding made of ALCOA aluminum. By the end of 1946, Hoess siding had been installed at housing projects in the northeastern U.S. including a 31-unit development just outside Pittsburgh. This subdivision was the first in the nation to make exclusive use of aluminum siding.[2]

ALCOA had some of the earliest aluminum-sided houses but lost interest in the 1950s, wanting to focus on primary ingot production. Reynolds would go on to dominate the aluminum siding market. ALCOA remained content to supply sheet and ingot to siding fabricators. In fact, ALCOA, in its monopoly days, had restricted its activity to being a supplier of fabricators. If it entered the market directly, it would go into competition with its own customers and actually lose business overall. Reynolds and Kaiser were free of such supplier arrangements and could vertically forward integrate into finished product. As the postwar demand for new houses diminished, manufacturers began promoting aluminum siding as the perfect material to cover old, outdated siding. Its use peaked in the 1970s at 22 percent of the market due in part to the 1963 introduction of vinyl siding. Vinyl did not dent, which had been a problem with aluminum. Aluminum remained a popular covering for farms from 1950 on.

Aluminum folding chair (Library of Congress).

Aluminum siding did become the mainstay of the mobile home. In particular, insulated modular panels became a key element in the production of mobile homes. The mobile home was the practical version of Bucky Fuller's Dymaxion Home merged with the utility of Airstream trailers. The original rationale for this type of housing was its mobility. Mobile homes were initially marketed primarily to people whose lifestyle required mobility. The mobile homes were built on chassis with wheels, clad in polished or enameled sheet aluminum with small sliding windows, and were sold completely finished with bathrooms, complete kitchens, wood paneling and furniture. However, beginning in the 1950s, the homes began to be marketed primarily as an inexpensive form of housing designed to be set up and left in a location for long periods of time, or even permanently installed with a masonry foundation. By the 1950s, hundreds of producers had sprung up across the country with names like Skyline, Fleetwood, and Champion, selling 10-by-60-foot mobile homes for $4,000. Mobile homes proved extremely successful, and by 1968 accounted for 25 percent of single family homes.

Trailer parks became common as the mobile home became a permanent form of housing. These trailer parks allowed homeowners to rent space on which to place a home. In addition to providing space, the site often provides basic utilities such as water, sewer, electricity, or natural gas, and other amenities such as garbage removal, community rooms, pools, and playgrounds. These parks became problems for many cities, who perceived them as cheap housing. There was still a demand for something very mobile.

Another unique product that utilized aluminum was the motor home. This was popularized by car camping trips. In the 1910s, people including Henry Ford himself started to extend living quarters onto the Model T Ford. For years, the self-styled "Four Vagabonds"—Thomas Edison, Henry Ford, Harvey Firestone, and naturalist John Burroughs—took camping trips around the nation. The trips were followed by a caravan of reporters who made the trips popular reading throughout the United States. The longer trips included the West Coast, Middle Atlantic States, and New England. They often added famous guests such as Presidents Calvin Coolidge, Herbert Hoover, and Warren Harding, or stopped to visit prominent men such as Luther Burbank in California. At a 1921 campsite around Hagerstown, Maryland, President Harding joined them for a dinner of lamb chops, ham, corn, potatoes, and biscuits. John Burroughs called Ford's food wagon the "Waldorf-Astoria on wheels." When President Harding died of a heart attack in 1923, Firestone, Edison, and Ford took a camping trip to the funeral in Marion, Ohio. The group then went to Edison's birth home at Milan, Ohio, and on to Ford's Dearborn, Michigan, estate before traveling to northern Michigan. Future president Colonel Dwight David Eisenhower joined them at the campfire in 1919 to discuss a national road system for the military.

This would be the beginning of the Recreational Vehicle (RV). The first true RV was Pierce-Arrow's Touring Landau, which came out in 1910. The Landau was more of a house car with the house being the back seat. The Landau had a back seat that folded into a bed, a chamber pot toilet, and a sink that folded down from the back of the front seat. In 1928 Pierce-Arrow introduced a more advanced house car, which looked like a school bus of today. Identified as the Privateer, it is one of only three models known to be produced on a Pierce-Arrow truck chassis before the market crash of 1929. Several other manufacturers tried to manufacture them in the 1930s, but the poor economy doomed their efforts. The 1963 Clark Cortez Motorhome was one of America's first front wheel drive motor homes with aluminum siding. They were built by Clark Equipment Company, which makes fork lifts and heavy construction equipment. Aluminum siding would create an RV boom in the 1960s. Again, the returning GIs and the Baby Boomer generation created a huge market.

Similarly, all things related to housing became a major market for aluminum. The appliance boom fueled by the housing boom offered another tonnage market for aluminum. At the end of World War II, large manufacturers had shifted their aircraft plants to produce modern labor-saving devices such as aluminum frame electric washing machines and dryers. Bucky Fuller and other 1920s pioneers had experimented with such appliances. These revolutionary consumer machines boasted the 1950s-chic white enamel finish. Consumer washer and dryer appliances were viable in large part due to the lightweight nature of their aluminum frames. The tubs were cast aluminum because of the corrosion resistance. Still, as with cars, steel was a tough competitor with aluminum frames. Of course, aluminum-framed appliances were at an advantage for use in mobile homes.

More heavy structural applications also evolved out of the development of aluminum alloys for siding and transportation. In 1933, the first aluminum bridge deck was used to replace an earlier steel and wood deck on Pittsburgh's Smithfield Street Bridge in order to increase its live-load carrying capacity. The rebuilding of the Smithfield Bridge eliminated 751 tons in the structure, which added an equivalent to the load capacity of the bridge. This new aluminum deck structure enabled the bridge to carry the new electrified trolley cars being introduced at the time in the city of Pittsburgh. It carried two lanes of motor traffic and two tracks for trolleys moving both directions. The Smithfield Street Bridge became the major artery of the time carrying such traffic across the Monongahela River into Pittsburgh.[3] The only alloy available at the time was the 2000 series, which would be a low-strength alloy today, but it had excellent corrosion properties.

The first all-aluminum bridge in the USA was constructed in 1946 for railroad traffic. It was a 100-foot single-track span railroad bridge. ALCOA constructed it on a line serving their Massena smelter as an illustration of the capability of aluminum in such applications. In 1950 the ARVIDA Bridge, a 290-foot arched aluminum bridge, was completed, spanning the old Saguenay gorge near the Shipshaw power house. This was built by Alcan. By the 1950s, new alloys in the 5000 series were developed for highway bridges. The first two of these were of relatively conventional built-up I-beam designs. These two-lane four-spans were in Iowa. Aluminum has since played a key role in many highway decks for bridges.

The biggest commercial (and cultural) success for aluminum after World War II would be the aluminum folding chair. The folding chair would come out of the product innovations after the war to use surplus aluminum. One big part of excess aluminum capacity was aluminum tubing of the 6000 alloy series. The folding chair used aluminum tubing for the frame and another product of the war, nylon, for the seat and back. This first aluminum folding chair was invented by Frederic Arnold in 1947. The folding chair was an almost immediate success. It proved portable enough to take to the backyard as well as to the beach. Historians describe its design success: "The design fit the multi-use, action-oriented suburban culture. During cold weather, the folding chair could be conveniently stored in the garage or basement. It was available everywhere and became the type of low end, inexpensive product that was sold in low-margin, high volume department stores."[4] It would become the most popular chair ever produced. By 1957, the Fredric Arnold Company of Brooklyn, New York, was manufacturing more than 14,000 chairs per day. It became the iconic symbol of 1950s suburban life.

Another result of high-strength aluminum used in the war was the aluminum ladder. Some aluminum ladders had been made of Duralumin in the 1920s, but the first commercial application by ALCOA was to supply an aluminum ladder to the Oslo, Norway, fire department in the mid–1930s. ALCOA was asked whether there was an alternative to the heavy and bulky timber extension ladders that they were using. They needed a type of ladder that was more reliable and easier to handle. Because it's light in weight, an aluminum ladder is ideal for commercial and domestic uses. Unlike a wooden ladder, it is not susceptible to fire. Aluminum does not rot or weaken the way wood does when exposed to the elements. This means it can be stored outdoors with no problem. However, ALCOA turned down the development project. Sam Cabris, the ALCOA engineer on the project review, resigned from ALCOA and created the Aluminum Ladder Company. With one customer, a small amount of money, and three employees, Sam Cabris rented a small building in Tarentum, Pennsylvania (not far from the ALCOA research center), to begin manufacturing aluminum ladders. The Aluminum Ladder Company struggled through the depression and the war to see the market for aluminum ladders take off with the postwar housing boom.

While construction was a high-tonnage application, art-related applications were still a growth area. Like the 1930s, there was a development of applications in art and household products in the 1950s. Fuller, Byam, and others' earlier projects and their close ties to the art world did open many to new uses of aluminum in the 1940s and 1950s. It seemed natural that colored anodized aluminum would surface from the merger of artists and engineers. First patent for protecting aluminum and its alloys from corrosion by means of an anodic treatment was in 1923. Anodizing was first used on an industrial scale in 1923 to protect Duralumin seaplane parts from corrosion. Aluminum alloys are anodized to increase corrosion resistance and to allow coloring, improved lubrication, or improved adhesion. The anodized aluminum surface is harder than pure aluminum, second only to diamonds with respect to its hard crystalline structure. Anodized aluminum extrusion products have a protective layer, and they are more resistant to wear from normal handling and usage. Anodizing paved the way for bright coloring of aluminum products. Colored anodized tube furniture became very popular in the high-end market of the 1930s.

In 1936, Dr. V. Caboni of Italy invented the famous coloring method consisting of two sequential processes: anodization in sulfuric acid, followed by the application of an alternating current in a metal salt solution. The color finish added to anodized aluminum is more enduring because the surface obtains more adhesive and porous qualities during the anodizing process. The resulting anodic film coating allows for effective dyeing processes to be applied. In the 1950s, colored aluminum was achieved by adding dyes. Colors of red, blue, and green often faded and appeared blotchy. Colors of gold, brown, gray, and black, however, usually retained their original brightness. Today colored coatings are produced by varying alloy content, which results in color on the surface only during the anodizing process.

Colored tumblers were the perfect complement to the suburban picnic of the 1950s. Advertisements often showed this colorful tableware with folding chairs. The first tumblers were deep drawn, anodized, and colored. The colored tumbler, like the aluminum chair, lacked design features such as durability. The product was often poorly designed from an engineering standpoint. They dented and scratched easily. Still, it was a product for the times. Their collectability has increased as iconic vintage items of the 1950s.

Cast aluminum cookware became popular in the 1950s. In 1950, Wear-Ever introduced NAD (New Anodized Design) heavyweight pans with anodized covers and welded lamp wedge-lock handles. It was a breakthrough in cookware. Anodized aluminum made the cooking performance of cast aluminum similar to that of cast iron in even heat distribution. The anodized surface also eliminated the leaching of aluminum into the food. Anodized aluminum has a naturally occurring layer of inert aluminum oxide thickened by an electrolytic process to create a surface that is hard and nonreactive. It has natural nonstick properties as well. Anodized aluminum continues its popularity even today.

Aluminum started showing up everywhere in the 1950s. Aluminum in the 1950s looked to replace wood in sports equipment like it had with folding chairs. In sports equipment, engineering design features were critical. On paper, aluminum's stiffness and modulus of elasticity offered a superior material for bows, skis, and tennis rackets. The general evolution in sports equipment went from wood to steel to aluminum to titanium to composites. Aluminum had shown success in archery equipment early on. In 1939, James Easton made the first arrow shaft out of aluminum. Aluminum offered a very straight, rigid arrow which became the most popular arrow all the way into the 1990s. This rigidity and stiffness led designers to look to other sports applications. The earliest work on aluminum skis had started in the late 1920s in France. In 1950, aluminum was first used in skis by American skier Howard Head. Head had the studied work of aircraft engineers in the late 1940s. The designers were three aircraft engineers—Wayne Pierce, David Richey, and Arthur Hunt. The light metal was sandwiched around a wooden core and welded by glue and heat, but this aluminum underside froze easily, increasing friction and stickiness. This negated its advantages. Howard Head modified the design and found success through the 1950s until aluminum was replaced by composite materials in the 1960s. Aluminum ski poles came out in 1955 and still have a market today.

Howard Head believed strongly that the properties of aluminum had many sports applications. Aluminum's stiffness made it ideal for tennis rackets, where that stiffness translated into more energy being transferred to the ball. An aluminum racket had been developed in France and appeared in the 1967 U.S. Open. In 1974, Wilson Sports purchased the rights to the design, and that year Jimmy Conners won the U.S. Open using a Wilson aluminum T2000 racket. In 1976, Head introduced the first American-designed aluminum tennis racket for the commercial market. Aluminum rackets gained a reputation for better control than wood or steel, and sales boomed. Head had taken advantage of the lack of rules by patenting his aluminum racket with an oversized frame. The oversized frame expanded the sweet spot for the average player. The other important property of a racket is its bending stiffness, which imparts energy on the collision of the ball and the racket. A flexible racket will bend more on impact, whereas a stiff racket of aluminum alloy will bend little and transfer energy to the ball. Flexible wood rackets, therefore, absorb more of the energy from impact, with more of the energy going into bending of the material. In comparison, stiff aluminum rackets are more powerful, as less energy is lost in frame bending and consequent vibrations, and more energy is returned to the ball via the strings. Aluminum rackets held the high-end market until being replaced by titanium and graphite composites.

The most iconic product of the 1950s was the aluminum jet commercial airliner. The first commercial jet was the aluminum British de Havilland Comet in 1953. The 36-seat Comet flew at 480 miles per hour. The top cruising speed of the DC-3 piston aircraft, in comparison, was about 180 miles per hour. Unfortunately, the Comet was the victim of a number of deadly accidents, and British Airlines suspended flights by 1955. Engineers found that the planes suffered from metal fatigue, especially around rivet holes, due to the cyclic pressurizing and depressurizing of the aircraft. This type of cyclic stress created small cracks that soon opened to catastrophic failure, which was typical of aluminum alloys. Metallurgists studying these early accidents found the rivet holes and sharp window corners where these cracks started. The solution was to eliminate these sharp notches and to have timely inspections of the rivet holes for these micro cracks. The Comet was redesigned based on the metallurgical analysis.

In 1952, Pan American Airways had already put in an order for the new 76-seat Comet 3, but the crashes put the order on hold. Juan Trippe, Pam American's legendary chief executive officer, had early on expressed a keen interest in operating a passenger jet service capable of flying nonstop across the North Atlantic. Having seen the bright promise of the British Comet and its failure, Trippe went to the biggest domestic airplane builders, Boeing and Douglas. The aluminum Boeing 707 and DC-8, respectively, were designed to meet the needs of Pan American and assure there would be no more fatigue failures.

In 1955, Trippe signed contracts with both companies to buy 45 of these jets (20 707s and 25 DC-8s). On October 26, 1958, amid much fanfare, Pan American inaugurated its New York–London route. On the very first flight, there were 111 passengers, the largest number ever to board a single regularly scheduled flight. Coach fares were $272, the same as one would expect to pay for a propeller engine flight across the Atlantic. British Overseas Airlines countered with the new de Havilland Comet 4, which incorporated improvements to remedy the problems with the older Comet 1. The British de Havilland Comet record was not the only problem; the Boeing 707 had accommodations for approximately four times as many passengers as the original Comet I. It was also 100 mph faster than the Comet. Aluminum had built a future for international air travel.

The most popular aluminum product of the 1950s was not in transportation, sports, or space, but in the simple TV dinner which is now part of the collection of the Smithsonian Museum. The Swanson TV dinner in its aluminum tray at a price of 98 cents helped change the very culture of the family. It was part of the social movement to free women from hours in the kitchen. Swanson Foods was a nationally recognized producer of canned and frozen poultry. In

1954, the company combined some new freezing techniques and the segmented aluminum tray used on airlines. The aluminum tray had turkey, potatoes, and a vegetable which was purchased frozen. It could be cooked in the oven in 20 minutes. It then could be put on an aluminum-legged folding TV table, introduced a year earlier. In 1954, ten million turkey dinners were sold.

While new products and processes highlighted the 1950s, the aluminum industry had its financial problems. Aluminum had become a competitive and strategic international industry. In 1958, finance problems led to British Aluminium's being taken over by Reynolds Metals and Tube Investment Group of England. The battle for British Aluminium began in 1950.[5] The takeover battle was the first aluminum war and it raged for a full decade. It was characterized by scandals, banking interference, and national politics. It would be Britain's first international hostile takeover.

The nature of the demise of British Aluminium is still debated. Between 1915 and 1953, British Aluminium spent around 18 years under direct government controls; and during both World Wars, a number of the company's senior managers were from or would go to the war ministries. It never transitioned well from World War II. The government had favored Alcan with some major war plants in Britain, which, coupled with Alcan's Canadian smelting advantage, made it much more competitive in England. British Aluminium lacked cheap smelting, production capacity, and well-developed bauxite mines. It had ownership of some potentially rich Australian mines but lacked capital to develop them. For a few years after the war, the company had cheap aluminum available from downed German aircraft. Still, British Aluminium was the sales leader in England in 1950, but Alcan had a cost advantage. British Aluminium had to sell at a cost loss to hold market share. The world oversupply of aluminum had strained British Aluminium to the limit financially by 1950. British Aluminium, with its old government structure, lacked innovation and creativity in developing new products. British Aluminium's initial strategy was to have ALCOA invest and partner with them in product development and sales.

The problems and the effort to have ALCOA buy British Aluminium became public in 1958. ALCOA would back out of negotiations for a period. The struggle for control of British Aluminium was now between ALCOA and Reynolds. ALCOA failed to be aggressive, and Reynolds pulled off a very successful partnership with British bankers. With its takeover of British Aluminium, Reynolds was now a leading international aluminum company along with Alcan. It ended the story of ALCOA as the whole aluminum industry. Now history had to focus on the world aluminum industry. There were six to ten major aluminum companies in the world, and the profits of the 1960s would bring even more new companies.

CHAPTER 17

The 1960s and 1970s—
The Aluminum Age

THE 1960S SAW THE DREAMS OF THE 1930S come to full reality. The 1960s saw new and old major markets for aluminum mushroom. After decades, many of Bucky Fuller's aluminum housing features became mainstream in the 1960s. The suburban picnic aluminum items continued their sales success into the 1960s. There were new headlines for aluminum from the emerging space program. The 1960s saw the development of the aluminum beverage can, which Reynolds pioneered through its research efforts. ALCOA also strengthened its research efforts and would perfect the aluminum can. Reynolds went another direction and developed new foil products that made aluminum foil a necessity in American kitchens. Aluminum siding for houses became common. Both Reynolds and Kaiser Aluminum reduced their power plants and smelting, focusing on product development to fuel the market boom. New problems would arise in the mining of ore and smelting from an emerging environmental movement. The environmental movement would also lead to a boom in the usage of aluminum cans. Still, not everything about the environmental movement was friendly to the industry.

More than the other domestic companies, ALCOA was forced to reverse a 70-year-old strategy of ingot and semi-finished goods. ALCOA was forced to move into the finished goods market to maintain high profits. The real profits were in the value-added products. ALCOA, which for decades had created hundreds of companies to use their ingot, sheet, and bars, now had to take on a strategy of acquisition of fabricating companies and internal development of fabricating divisions. It was a balancing act for ALCOA, since for decades its largest customers were fabricators. ALCOA moved into siding, telephone booths, wire, food packaging, foil, and other finished products. The effort was met with a mix of success and failure. In addition, the government reopened antitrust investigations on many of ALCOA's acquisitions. ALCOA also had lost its price leadership role and was facing worldwide competition. Kaiser and Reynolds seemed to be favored by the government.

During the 1960s, the aluminum industry became truly international. The North American oligopoly (ALCOA, Alcan, Kaiser and Reynolds) lost control of pricing and market. Most of the world's continents would have their own aluminum industries. Profits would continue to improve and aluminum would find new large markets. Still, 61 percent of the non-communist production was controlled by the Big Six, with ALCOA at 1,700,000 tons capacity, Alcan at 1,580,000 tons, Reynolds at 1,270,000 tons, Kaiser at 1,020,000 tons, Perchiney (France) at 886,000 tons, and Alusuisse (Switzerland) at 476,000 tons. This global market and production would bring new challenges. ALCOA would remain the senior company, but Reynolds, Kaiser, and Alcan would also continue to grow. New millionaires and billionaires would be created in the industry and new companies would emerge. These new aluminum companies in America included Anaconda, Martin Marietta, National Southwire, National Steel, and Revere. Competing materials such as high-strength steel, titanium, composites, and magnesium would challenge some traditional markets. Aluminum, not steel, would become the metal of economic and military strength.

The American aluminum industry was now international; it depended on imports of bauxite from foreign countries. The traditional attitude of ALCOA and the industry towards tariffs would make a major change from supporting them in order to protect domestic industry, to favoring open and tariff-free trade. The General Agreement on Tariffs and Trade (GATT) and the International Monetary Fund would evolve over the decade of the 1960s. It would virtually eliminate tariffs on bauxite, creating new profits for the international aluminum companies. However, America would also be flooded with cars, steel, tool machines, clothing, and furniture that, in the long run, brought in finished aluminum. Still, this type of global manufacturing, trade, and business was well suited for the aluminum industry. Aluminum was one of the few industries that depended on an international supply chain and market. ALCOA, however, found free trade to be a mixed blessing in the long run. The American aluminum industry would be in the rare position of growing under the free trade movement, although the wolf would come in the 1980s. The new postwar world would create new challenges in the 1970s.

Aluminum companies would be impacted by the 1970s energy crisis, since energy was the primary raw material of aluminum making. Energy had defined the aluminum industry since its conception. In October 1973, U.S. support of the Israeli military in the Yom Kippur War created a political backlash in the Arab nations. Arab oil-producing countries imposed an oil embargo on the

United States and increased prices to its European allies by 70 percent. Oil jumped from $3 a barrel to over more than $5 a barrel overnight. By the end of the embargo, it had reached $11 a barrel. In 1973, the United States relied on foreign countries for 36 percent of its oil. The embargo caused consumer panics, with stations running out of gas and mile-long lines of cars waiting to refuel. ALCOA and most American smelters saw power costs increase 400 percent. Electricity costs typically constituted 20 to 30 percent of production costs. Profits dropped dramatically by 1975 for American aluminum companies.

The run-up of oil prices created a time of hyperinflation in the late 1970s. Inflation did increase metal prices significantly, but it brought interest rate problems. The period from 1972 to 1982 saw ALCOA's revenues increase from $1.8 billion to $4.8 billion. The problem was that operating profit as a percent of sales dropped to that of the 1960s. High interests prevented ALCOA and others from expanding and improving operations. In general, the industry struggled but fared better than most businesses in the energy and inflation crises of the 1970s. Having met the financial challenges of the 1970s, new issues came to the forefront late in the decade.

The 1960s and 1970s brought a new type of criticism. Environmentalists focused on the dark side of alumina mining. While ALCOA was proud of its raising of the standard of living in their mining areas, there were other issues. The Surinam mine at the time was ALCOA's largest mine. In 1966, a hydroelectric plant and smelter was started there to reduce the high cost of shipping alumina ore to North America. The project required a major dam on the Surinam River. The dam and flooding would cause the dislocation of 6,000 people from their homes. Up to 43 small villages and related cultural sites were destroyed. The dislocated included a population of indigenous people and maroon (a mix of indigenous and former black slaves).[1] The new villages were criticized for lack of basic utilities and because no property rights were given for the lost land. Colonial capitalism came to be seen as evil for many developing companies and an easy target for revolutionaries. The press attack on environmental sins was something new for ALCOA but would be part of its future. ALCOA and other aluminum companies would have to learn how to better handle such mining problems, which had been overlooked for decades. While vertically integrated, ALCOA struggled with environmental issues within its mining operations; Reynolds and Kaiser were able to focus on product development.

Reynolds Metals was a pioneer in research and development of new products using aluminum in the 1960s. These included an aluminum bus and other aluminum motor vehicles. For Reynolds, the biggest breakthrough was the development of powerful foil-rolling machines to make thinner foil. These machines made it possible to roll a pound of aluminum into foil that would cover 480 square feet. Reynolds became the most product-oriented of the American companies. Reynolds's real target was the American kitchen. In 1965, almost 400 million pounds of aluminum foil were produced in America as millions of housewives found it indispensable in the kitchen. Reynolds moved into high-tech products, like an aluminum submarine, which had only existed in science fiction.

The concept of an aluminum submarine was developed at Reynolds during World War II by Louis Reynolds, who was in charge of the foil division. The idea was never fully developed during the war. In 1964, Reynolds had the Electric Boat Division build the world's first aluminum submarine and reintroduce aluminum to marine applications. The sub was called *Aluminaut*. It weighed 80 tons, was 51 feet long, and could accommodate a crew of 3, plus 3 to 4 scientists. It had 4 viewports, sonar, manipulators, and could hold 6,000 pounds of payload. Reynolds had the *Aluminaut* designed and built as an experiment for ocean research. For flexibility, it was outfitted for many types of oceanographic and salvage missions. The *Aluminaut* became famous in its search for the lost H-bomb off Spain. Since aluminum's strength-to-weight ratio exceeds that of steel, the *Aluminaut*'s 6.5-inch-thick shell could withstand pressures of 7,500 pounds per square inch. Kaiser Aluminum followed with its own sub, *Deep Quest*. In 1964, Reynolds produced another aluminum research sub for the U.S. Navy called *Alvin*. In the 1970s, the hull was replaced by titanium. *Alvin* became famous for its work with finding the sunken wreck of the *Titanic*. In 1969, the *Queen Elizabeth 2* used an aluminum superstructure and a great deal of aluminum cladding.

The 1970s would bring the true fulfillment of Jules Verne's dream of an aluminum-hulled ship. In the 1930s, aluminum had been used in many nonstructural marine applications such as stairs and decks. The first oceangoing aluminum-hulled ship would be for short transport. The *Sacal Borincano* was an all-aluminum ship used to carry forty highway freight trailers between Miami, Florida, and Puerto Rico. Smaller aluminum-hulled ships for the military and the Coast Guard were also built. There was a fear of metal fatigue cracking, which was common in aircraft aluminum, and the best use seemed to be not in the superstructure.

The British built a few frigates with aluminum superstructures in the 1970s. This was based on the Japanese success in the 1950s of all-aluminum Coast Guard patrol boats. These were the Type 21 frigates designed for speed and fitted with aluminum superstructures. The superstructures became famous during the Falklands War. Two smaller ships in the Falklands, the frigates

Antelope and *Ardent,* which were sunk after devastating fires, had aluminum superstructures and steel hulls. Each of the frigates contained about 800 tons of steel and about 130 tons of aluminum. Initial analysis raised some question, still pondered today, as to whether the low melting point of aluminum during fires from missile attacks contributed to the sinkings. In the Mediterranean Sea in November 1975, the American guided missile frigate USS *Belknap,* equipped with an aluminum superstructure, collided with the aircraft carrier *John F. Kennedy* during a night exercise. The frigate caught fire, and the forward five-inch gun turret fell through the melting aluminum deck.[2] Eight sailors died and 48 others were injured on the USS *Belknap.* Both ships caught fire, but it was the *Belknap,* with its aluminum superstructure, that nearly perished in the subsequent 2½-hour inferno. Sailors on the *Belknap* reported seeing the aluminum actually burning. The *Kennedy*'s fire was out in ten minutes.

As these aluminum superstructure frigates remained in service for extended periods, a new problem arose. Fatigue-induced cracking of aluminum superstructures remained a serious problem with frigates still in service. Cracking is usually caused by a combination of applied cyclic stresses and stress concentration interacting with a region of material weakness such as a weld. In the 1990s, the Navy quit building aluminum-superstructure warships. Their main reason was repairs resulting from fatigue. Repairs to ships were much more difficult than those to aircraft. While the use of aluminum in ships increases annually, the true aluminum-hulled warship has yet to be built.

Aluminum trains, once an experiment of the 1930s, would make major inroads in the 1950s and 1960s as applications for aluminum were actively studied. The Toronto Subway opened in 1954 and one experimental train set consisted of the first aluminum subway cars, which reduced weight and, therefore, operating costs. In 1963, only aluminum was used for the next car order; and as the new cars measured 75 feet at the time, they were the longest in the world. Subway aluminum cars turned into a long-term success because they did not compete with the airlines. Aluminum trains were another story; they required great speed to successfully compete with the airlines.

The aluminum TurboTrain built in 1969 was lighter, faster, quieter, smoother, and more reliable than conventional trains—and cheaper to run. The TurboTrain was the hope of the railroad industry to take market from the airlines. Conceived on aerodynamic principles and powered by aircraft-type gas turbine engines, the TurboTrain was designed by United Aircraft Corporation. It was built in lengths ranging from three to nine cars per train and had a streamlined, airplane-like exterior designed to minimize drag. It was reminiscent of the 1930s, with a skin of smooth, heavy-gauge aluminum. The Turbo-Train could travel at speeds up to 170 mph, but initial top speeds in passenger-carrying service were about 100 mph. Penn Central operated TurboTrains between Boston and New York as part of the Northeast Corridor high-speed ground transportation project under contract to the U.S. Department of Transportation. In Canada, Canadian National Railways operated the TurboTrains between Toronto and Montreal.

Penn Central's 3-car sets carried 144 people and operated at a maximum speed of 100 miles per hour. In the U.S., they were decommissioned by 1972, and by 1982 in Canada, because TurboTrains employed many advanced and derived technologies, which led to high maintenance costs. The oil crisis of the 1970s also made the turbines too expensive to operate even with the reduced weight of aluminum. Aluminum would, however, continue to make progress in the freight car market. Today more than 200,000 aluminum freight cars are in service, and more than 45 percent of new rail cars are made of all or mostly aluminum construction. In the 1960s, the real breakthrough of aluminum wouldn't come from transportation, but rather from packaging.

Chapter 18

"Tin Cans" and Space Capsules

THE 1960S WOULD BRING ALUMINUM'S biggest new application in the form of the aluminum can. The first steel (tinplate) beverage can was introduced in 1935 by the Krueger Brewing Company. The early three-piece steel cans had a flat top that required a "church key" for opening. Cone-top capped steel beer cans also appeared in 1935 and remained in production until about 1960. These were developed by Crown Cork & Seal, a leading beverage packaging and beverage can producer. The three-piece can was created with a top, bottom, and body piece. The body piece was a sheet formed into a cylinder and then welded. The top and bottom were then welded to the formed cylinder body. By the early 1950s, the brewing industry was becoming concerned about the growing cost of tinplate or tin-coated steel, prompting Kaiser Aluminum to establish a container R&D facility in Chicago to develop an aluminum can. Aluminum was a natural container material. Tinplate cans could rust in long storage or corrosive environments, while aluminum offered total protection. Aluminum's lighter weight offered a reduction in transportation costs. The first generation of aluminum cans weighed approximately three ounces per unit versus four for tinplate.

In neutral Switzerland, tinplate had been almost unobtainable during the war. Consultant and Swiss engineer Jacob Keller heard that can makers were struggling, so he adapted a press with ironing dies to form a container from a disc of soft pure aluminium. Keller took out patents in 1946. His invention looked promising, and Alcan started to fund Keller's research. But as tinplate again became available, interest declined.

In 1954, the Adolph Coors Co. joined forces with Beatrice Foods to found Aluminum International, Inc., and develop an economic process for manufacturing aluminum cans for their products. Their first eight ounce cans were marketed in 1958 by Beatrice Food's Hawaii Brewing Corp. for their Primo beer. Coors beer followed in 1959. The eight-ounce can was short of the 12-

ounce standard for beer, the properties of aluminum proved the limiting factor in processing. Coors headed up the R&D efforts; they developed a method of continuously casting aluminum and rolling it to about ⅛-inch thickness, then blanking circular slugs that were the diameter of the can. The new approach greatly increased the production rate. This early technique for making cans was slow and plagued with tooling problems, which continued to limit the size to eight ounces. At the same time as Coors, Kaiser Aluminum set up a development line in 1957, but dropped out when market research showed canners would resist changing the tooling from steel to aluminum.

Reynolds's research effort proved more successful. Reynolds adapted the "draw and iron" (D&I) process of Jacob Keller to produce aluminum cans, and began shipping seven-ounce beer cans in 1958. Reynolds invested heavily in developing an improved process. The Reynolds Metals Company had initiated its own specific R&D program for manufacturing a 12-ounce aluminum can, and in 1963 pioneered the production method for economically producing 12-ounce aluminum cans that are used today. The first 12-ounce aluminum beverage can was manufactured by Reynolds Metals Company in 1963 and used to package a diet cola called Slenderella. Royal Crown adopted the aluminum can in 1964; and by 1967, Pepsi and Coke followed.

These new aluminum cans were two-piece cans. The body was formed by drawing out a circular blank with a punch into a one-piece cylinder, only requiring a lid. The two-piece construction greatly reduced overall production costs. The aluminum can soon became competitive with steel cans for beer. In 1963, Reynolds started making aluminum cans for Hamm's Beer, while ALCOA was making them for Budweiser.

The draw and iron process removed all other production methods from the aluminum can industry, improved the life of machinery, and created lighter cans. While other industries had used similar techniques, the Reynolds aluminum can was the first time such a process was used for beverage cans. The use of aluminum also led to a new top opening not requiring a church key. For years the church key was a necessity for all steel cans. Aluminum cans would use a pull-top opener. The early Coors cans used a pushbutton top, which was an improvement over the church key.

Initially the pull-top can design in 1963 was licensed to ALCOA and Pittsburgh Brewing Company, the latter of which first introduced the design on Iron City Beer cans. Iron City used a three-piece can with the new aluminum pull-top and aluminum body. The aluminum pull-top was soon being used on steel cans and was a major aluminum market in its own right. The failure rate of early pull-tops frustrated consumers. When it worked right, it detached easily, but the problem was what to do with it. The environmental movement was underway, and pull-tops were a major source of litter at beaches, parks, and street corners.[1] Environmentally conscious consumers had one solution. The removed tabs were then put back into the drink cans, causing a danger of swallowing them. The American Medical Association noted cases of children ingesting pull-tabs that had been dropped into the can. In 1975, Daniel F. Cudzik, an engineer with Reynolds Metals, filed a design patent application for a new tab opening called a stay-on tab. This later became known as a "Sta-Tab." The Sta-Tab required the development of a new alloy by ALCOA. The can lid was made of the alloy 5182 containing 4 percent magnesium and 1 percent manganese, which is stronger and harder to form than the can-body alloy. When the Sta-Tab launched in 1975 on Falls City Beer, and quickly on other beverages, there was an initial period of consumer testing and education; but it was the standard by 1980. The next innovation in the aluminum cans would be a lighter, thin-walled can.

The Can Division of Reynolds did a commercial production evaluation run of a thin-wall can in 1970 using a new alloy and process. This new process still formed the can using a punch, but did not require softening, anneals, or stress reliefs. It would punch and form the can in four quick steps. It would also use a specific aluminum alloy not yet in use in beverage can manufacturing. The alloys used contained 92 percent to 97 percent aluminum, 5.5 percent magnesium, 1.6 percent manganese, 0.15 percent chromium, and some trace amounts of iron, silicon, and copper. Alloys used today include 3004, 3105, or other 3xxx/5xxx series aluminum. In the early 1970s, ALCOA introduced its Featherlite can that was drawn from light-gauge aluminum stock and was 20 percent lighter than the average aluminum can in 1972, resulting in major transportation cost reduction. The thin-wall can allowed for 30 cans per pound of aluminum versus 20 cans. The reduction in material costs was another offset for the cost of tooling for aluminum. The wall thickness went to five-thousands of an inch, or less than the thickness of two pieces of paper. As a result, the production of aluminum cans went from 100 million cans in 1970 to over 2 billion in 1974.

Canning industry conversion was still slow due to the initial high cost of tooling needed for aluminum. By 1990, 97 percent of all beer and soft drink cans were made of aluminum. Reynolds and Kaiser moved directly into the can market, while ALCOA preferred to increase its sheet production and work with can makers.

Recycling of aluminum cans became another advantage, as can makers who had major tooling costs in moving to aluminum, created recycling programs. The environmental movement and the government pushed for more consumer recycling. In 1965, the Solid Waste Disposal Act passed by Congress started an industry-wide research effort into recyclable products. By the late '60s, bottle deposit laws started to require recycling. Aluminum cans with 100-plus years of life presented an environmental issue. Recognizing the value of used aluminum cans as a raw material for making new cans, the aluminum industry begin creating a massive system for recycling and redeeming used beverage containers. There was also competitive pressure from steel cans that would rust away in garbage dumps. The steel industry in the 1980s countered with a two-piece can of their own. Steelmakers had also developed a thin-wall drawn and ironed can to compete with aluminum. The aluminum industry needed to back recycling programs to maintain their inroads into can manufacture.

Recycling was not new, but an extensive collection network was needed. The Coors Company began a successful program called Cash for Cans in 1964, offering one penny for each can returned for recycling. By 1966, Coors was honored by the Good Outdoor Manners Association for environmental stewardship, having collected over 13 million cans in the previous year. The aluminum industry was able to build a very successful collection network. U.S. collection grew from 1.2 billion cans in 1972 to more than 62 billion cans in 1995 through curbside recycling programs and more than 10,000 recycling centers. Today, aluminum cans are the most recycled packaging in the world. Scrap aluminum requires only 5 percent of the energy used to make new aluminum, making it good business to recycle. For this reason, approximately 31 percent of all aluminum produced in the United States today comes from recycled scrap. Aluminum successfully caught the favor of the green movement as well by the 1980s. The total recycling rate for all aluminum products today is around 65 percent in the United States.

Steel cans did make a small comeback with their own technology improvements. The ease of aluminum recycling became a problem for steel cans. Aluminum cans also had many cost advantages over steel by 1980. Today aluminum holds about 75 percent of the total can market. Tinplate is still used in acidic liquids such as tomato juice and fruits. Aluminum cans continued to grow as a market and represented one of the great successes of the aluminum industry in the 1960s and 1970s.

Ever since the launch of *Sputnik* in 1957, aluminum has been the material of choice for space structures of all types. Chosen for its light weight and its ability to withstand the stresses that occur during launch and operation in space, aluminum has been used on *Apollo* spacecraft, the *Skylab*, the space shuttles, and the International Space Station. The cover of the first satellites launched in the 1950s was made of aluminum. From one-twentieth to half the weight of the rocket launcher, and up to 90 percent of the space shuttle, is made of aluminum. In addition, powdered aluminum fuel engines were used for the launch of space vehicles in the 1960s and 1970s. Powdered aluminum fuel actually evolved out of the space program. Unlike rocket frames, aluminum was not a natural for fuel. Early rockets from Goddard to the German V1 and V2 used liquid fuels.

After the war, the push for more reliable fuels for long-range missiles and space rockets started to look at solid fuels. The first successful solid aluminum propellant ballistic missile was the U.S. Navy's Polaris A1 submarine-launched missile in 1960. By the mid–1960s, the first Air Force solid propellant ballistic missile was the Minuteman, so named because it could be ready to fire on a minute's notice and did not need to be liquid fueled before launch. The first space launches used aluminum solid propellants. Most important was the building of the aluminum superstructure, the Vehicle Assembly Building at Cape Kennedy. This superstructure had more space than any building at the time. It would be the iconic symbol of American space superiority.

One of the most popular aluminum space products has been the space blanket. Neil Armstrong's space suit was made of space-blanket material that included a layer of pure gold. NASA manufacturers created the material by depositing vaporized aluminum onto a very thin plastic film. The resulting material is thin, flexible, and heat-reflective. It is very light in order to minimize weight impact on vehicle payload while also protecting spacecraft, equipment, and personnel from the extreme temperature fluctuations of space. Early in the space program, the National Metallizing Division of Standard Packaging Corporation was a supplier of this reflective material to NASA. In fact, it was one of the original subcontractors NASA turned to for design and supply of the material. The thin blanket version can be traced to the NASA program of the 1970s.

It was National Metallizing Company that NASA turned to for assistance when, in May 1973, during the first few days that *Skylab* was in orbit, it was malfunctioning and overheating. A heat shield broke off during launch, and air temperature inside the orbiting station began approaching 130°F. NASA was concerned about the condition of food, film, and other equipment inside, as well as plastic insulation and possible toxic gases if the temperature rose too high. The staff at National Metallizing and NASA was called on at Marshall Space Flight Center to help create an emergency parasol-type sunshield that helped save millions of dollars' worth of equipment and years of research with the *Skylab*.

Chapter 19

Space Age Products

THE RACE TO THE MOON HAD GIVEN AMERICA and the world a new basket of super materials from silicon chips to space blankets. Most importantly, the use of aluminum changed its utilitarian image back to a more exotic one. Markets even reopened in housing, automotive, and giftware. The properties of lightness, strength, heat control, reflectivity, and ease of fabrication, all of which made it popular with NASA, also opened old and new doors with the consumers.

Thin aluminum-coated space blankets seemed miraculous to the consumer. These blankets slow sweat evaporation and block wind; they reflect 90 percent of body heat, minimizing cooling through convection and radiation. Place the shiny side next to your body, as the dull silver side reflects only 65 percent of radiated heat. Space blankets are often given to marathoners and other endurance athletes at the end of races, or while waiting before races, if the weather is chilly.

The 1960s brought a new popularity in aluminum siding as a space age material, but it started to lose market share to vinyl by the early 1970s. In the first 15 years after World War II, aluminum clapboard siding was applied to more than three million homes in the U.S. In the 1960s, it was marketed to consumers as a remodeling product for older homes. By the 1960s, aluminum had captured fully one-quarter of the residential re-siding market.

Despite aluminum's popularity with consumers, vinyl quickly took market share in the 1970s. While aluminum siding declined, aluminum gutters boomed. In the 1960s, seamless aluminum gutter machines were invented, which drastically changed the way gutters were produced. Because of the strength and lighter weight of aluminum as compared to steel and other metals, aluminum gutters became popular. The invention of seamless aluminum gutters changed the situation completely. A machine was developed that holds a roll of aluminum, called gutter coil, at one end, and extrudes a formed gutter at the other end. The formed gutter can be cut to any length desired and custom fit to the home on site. This method of creating gutters without seams made the gutters stronger and leak-proof. Today, over 70 percent of residential gutters are seamless aluminum.

The idea of an aluminum house was also revived in the early 1960s with the space program. Aluminum was now a space age material, reviving its early status as a very special metal. In January of 1957, ALCOA announced the formation of a Residential Building Products Sales Division to manage its expansion into the homebuilding market with its aluminum house. While not a Dymaxion, it did use aluminum extensively. These were designed for middle-class families of four to six. To jump-start sales in this newly formed division as well as explore and encourage new uses for aluminum in homebuilding, ALCOA also announced it would be sponsoring the construction of 50 "Care-Free" aluminum model homes priced under $25,000. The company stated its aim was to create a lower-maintenance home and achieve the greatest change in residential building materials in centuries. Only 24 of the ALCOA "Care-Free" homes were completed across the U.S. Unfortunately for ALCOA, they never made it to the planned 50 model homes. The final cost of the homes was much higher than expected, ranging from $35K to $60K, which led to the project's eventual demise. The insulation attributes of aluminum products used in space capsules found consumer applications in housing. Many of the insulation features morphed into aluminum-framed windows, which did become popular in the 1960s.

The building of an aluminum mansion for the wealthy proved also to be a bridge too far. The grandson and heir to Alfred Hunt's fortune, also named Alfred Hunt, would do just that. Part of his inheritance included his grandfather's mansion, Elmhurst, the turn-of-the-century Pittsburgh mansion in which he grew up. The residence didn't fit his lifestyle as a bachelor and a visionary. He wanted to showcase the creative potential of aluminum. And so he hired Los Angeles architect Edward Grenzbach to create a new home. The exterior was built of slabs of both cast and extruded aluminum. The interior included brushed aluminum steps, an aluminum ceiling in the den, ornate chandeliers made from aluminum, aluminum moldings around the doors and windows, aluminum blinds, and even a 60-by-10-foot aluminum lap pool. A recent owner noted: "The house was very cold; brown walls and lots of aluminum, steel and glass furniture, those vertical aluminum blinds. I made an effort to warm it up with as much wood as possible."[1]

The late 1960s saw aluminum art and giftware emerge again as a novelty metal. Government space research would bring new art alloys for giftware. Wilton Brass Company, which initially produced industrial brass, aluminum, and iron castings, then moved into aluminum giftware. The company's greatest success occurred in 1963, when Wilton developed the formula for a unique aluminum-based alloy called Armetale, which led to a new line of Wilton products. Armetale is similar to Nambe, an eight-metal alloy whose major component is aluminum. Nambe was developed in 1953 by Martin Eden, a former metallurgist at the Los Alamos National Laboratory. Similar alloys appeared as the popularity of this alloy in gifts rode a wave into the 1970s and 1980s. Armetale, Nambe, and similar alloys have the luster of silver and the solidity of iron. These alloys do not contain silver, lead, or pewter, but have a heavy feel. They were sand cast and polished to a high luster to create products such as cookware, bowls, platters, trays, dishes, mugs, plates, napkin rings, candle holders, wine bottle holders, martini shakers, and coasters. In appearance, the alloys were "pewter-like" in look and feel; and most consumers would not consider it aluminum, which was its advantage. These alloys remain popular today.

Aluminum tried to make further inroads in sports in the 1960s and 1970s with its new image. "Space age" aluminum shafts caught the attention of the golfing public. The one man most responsible for the excitement was Arnold Palmer. With aluminum shafts, Palmer won the 1967 Los Angeles and Tucson Opens and finished high in the money in many of the others. It was quite natural that the Arnold Palmer Company would be the first in introducing this aluminum shaft to the general public. Spalding moved into aluminum shafts in 1969. The company's aluminum shafts appeared in the '70s but didn't take off, as they became brittle in cold weather and broke very easily. The big benefit was they were lighter than steel and didn't rust. They also had better stiffness than wood shafts to transmit more energy to the ball. The aluminum debate continued, but there were many golf industry criticisms. One executive said, "It is doubted at this time whether aluminum will supply the answer to those who are looking for a panacea to their problems. It may help the elderly and the ladies, but we're not sure it is going to help too many others. Nor will it improve the play of the tournament golfer. At least they are not clamoring for it yet."[2] With the advent of titanium shafts in the 1990s, aluminum as a golf shaft material was dropped.

Aluminum golf heads had a much earlier history. Spalding was selling aluminum fairway club heads by 1910, and its Gold Medal series (1910 to 1919) featured aluminum bronze. In 1907, aluminum alloys such as the 6000 series (mainly 6060) appeared. The popular space alloy 7075-6A Aluminum (aluminum + 4 percent to 6 percent lead + 2 percent to 3 percent magnesium) was used for drivers because of its light weight and strength. More recently, harder alloys have been developed for golf heads, but aluminum does not have much popularity in golf today.

Unfortunately, aluminum would suffer a major setback in the auto industry in the late 1960s. Aluminum in cars had made inroads in parts such as pistons in the 1920s. The aluminum body was still being held up on cost. The major cast iron part in the 1960s was the engine block. Cast iron was used because of its ability to be precision sand cast. Aluminum had all the heat transfer properties of cast iron with a number of advantages. Aluminum's principal advantage as an engine material over cast iron was weight. Aluminum weighs about one-third as much as iron and could save a couple of hundred pounds. Aluminum is also a better heat conductor than cast iron, so using it for the cylinder block of a water-cooled engine allows the use of a less-bulky water jacket, thus saving additional weight. All the research at universities switched from steel to aluminum for space applications. Spinoffs of these new aluminum products started to resurface in the auto industry. One of the biggest technology challenges in the auto industry was the cast aluminum engine blocks.

American Motors Corporation introduced the first aluminum block in the Dodge Rambler. The Rambler had an aluminum 196 six-cylinder engine in 1961. It was America's first die-cast aluminum block six. Aluminum was not totally new to engine blocks. Aircraft engines had aluminum blocks, and Europe had some success in cars in the late 1950s. In the early 1960s, Buick developed a cast aluminum V8 engine block. By 1963, GM had lost confidence in the aluminum V8. The aluminum engine's maintenance issues had given it a bad reputation, and the blocks were too expensive to manufacture. American Motors ended their production the same year for similar problems.

The aluminum engine required more stringent compliance with the maintenance schedule than the cast iron engine, but many owners just treated the engine as they would have treated a standard cast iron engine. However, if the engine coolant got low or the car overheated, there would be damage to the aluminum block. A cast iron block was much more forgiving, but many owners ran into problems with their aluminum engines. When an owner did experience difficulties with the aluminum engine, American Motors dealerships would replace the engine with a cast iron block. American Motors ended the experiment in 1966 as America was about to put aluminum on the moon. The inability of American engineers to solve the engine problems seemed strange; but it required, as the space program had, the combination of the best minds in the field to take up the challenge again.

The Chevy Vega was conceived in 1968 to utilize newly developed all-aluminum die-cast engine block technology—the first sand-cast aluminum blocks. This sand-cast technology brought the cost down to that of cast iron. The Chevy Vega was powered by an inline four-cylinder engine with a lightweight, aluminum alloy cylinder block in the early 1970s. The Vega aluminum engine was a joint effort by General Motors, Reynolds Metals, and Sealed Power Corporation. The lightweight Vega was to be America's challenge to small car foreign competition. It was to be built at Lordstown, Ohio, at a new plant. Production at Lordstown was projected at 100 Vegas an hour—one every 36 seconds—from the outset. Twice the normal volume, this was to be the fastest rate in the world, although it was never fully achieved. As the workers moved past the one-per-minute operation barrier, the human machine broke down. Furthermore, it would create serious labor issues at the plant, and this would contribute to overall quality problems with the Vega model. The car earned *Motor Trend*'s 1971 Car of the Year award based on its breakthrough aluminum engine. Interestingly, today the Vega often makes the "10 worst car" lists.

Perfecting the aluminum engine would prove more difficult than landing a man on the moon. The Vega model, and particularly the aluminum engine, tarnished the image of General Motors and aluminum. The Vega was a subcompact meant to meet the Japanese competition head-on even before the oil crisis of the 1970s. It had been marketed as the answer to Japanese imports. Vega sold in huge numbers. During its first model year, 1971, Chevrolet sold 277,700 of them. During 1972, out went another 394,592 units, then 395,792 in 1973 and 452,886 for 1974. The oil crisis pushed the sales total to over two million for its seven-year run. The Vega's aluminum engine block, which dramatically reduced weight, never performed well. The engine proved noisy and prone to overheating. Oil leaks were common as well. When the engine got hot, which happened often because of a poorly designed cooling system, the cylinders distorted, and the piston rings wore off the cylinder walls. Then, at best, the cars burned more oil. At worst, the distortion damaged the head gasket, caused the coolant to leak, and eventually destroyed the engine. It would make for a poor image of American small cars against the surging Japanese imports. This only compounded other quality problems.

Forty years later, *Popular Mechanics* summarized the damage to the auto industry from the Vega:

> The result was that literally hundreds of thousands of buyers were having awful experiences with the car. Some were merely disappointed. Many were incensed. And a lot of them felt betrayed by General Motors, Chevrolet, and the American auto industry as a whole. Surely those customers were then far more willing to consider the Japanese alternatives that were starting to arrive. By the end of the 1970s, the once-ubiquitous Vega was already disappearing from America's roads. With such a crummy reputation for reliability, the Vega's resale values soon dropped down near zero. Legend has it some salvage yard even put up "No Vegas" signs to announce that they weren't even bothering pulling usable parts off the cars before crushing them.[3]

One area where aluminum made progress in the auto industry during the race to the moon was aluminum cast and forged wheels. The use of cast aluminum wheels was premiered by Ettore Bugatti in 1924. These wheels incorporated an integrally cast brake drum, which amounted to a considerable weight saving, and Bugatti's wheels acted as large heat sinks to provide improved brake cooling. In the 1940s wheels were formed of two pieces of pressed steel, the rim and the disc, either welded or riveted into a single unit. With Bugatti's design, they were fabricated of a steel or aluminum rim connected to a center hub by metal spokes. A transitional design was a hybrid developed in Europe in 1949 utilizing a steel disc for strength and an aluminum rim for weight saving. Such a design was used by Porsche and Jaguar in the 1950s. Another example was the Borrani wheel in Italy. Some French companies such as Panhard advanced aluminum wheels in the 1950s.

In late 1954, Cadillac El Dorado introduced the Sabre-Spoke aluminum wheel, manufactured by Kelsey-Hayes. In 1960, Pontiac introduced their version of the aluminum integrated wheel and brake drum, and this was followed by Corvette and Ford Mustang in the mid–1960s. A set of cast aluminum wheels cost as much as $500 in 1964. Forging is extremely expensive, so wheel prices could be twice the price of a cast wheel. However, there were advantages; forged 6061 aluminum wheels offered good corrosion resistance and reduced the car's overall weight. Still, cast and forged aluminum wheels were a very expensive option in the 1960s. Forged aluminum commanded an even higher price, 25 to 30 percent lighter than cast rims. Forged rims were impact-resistant and durable, which are musts for reducing the total weight of a car. The oil crisis made aluminum wheels available on most cars in Europe in the 1970s, followed by America in the 1980s. Europe would lead once again with the forged aluminum wheels versus cast.

CHAPTER 20

The Best and Worst of Times, 1970–2000

By the early 1970s, aluminum's image had been tarnished. It was now commonplace and viewed as cheap. The Chevy Vega engine had proved that everything aluminum was not always better. Things were only to get worse in the 1980s, yet aluminum usage continued to grow in products like cans. Aluminum, once the metal of the rich, had become a symbol of the lower classes by the end of the 1970s. The metal that was once reserved for royalty was now the metal of the proletariat. For the most part, this didn't reduce corporate profits; however, aluminum switched to being a commodity like steel and iron. Aluminum as the high-tech metal was replaced by titanium, and later by composites. High-price aluminum tableware would be resisted by consumer sentiment. Aluminum had lost its shine with the public. Cast iron, the metal of the Victorian era, had not seen the same decline as a material. Some had seen the decline of aluminum siding as a change in how the public viewed aluminum. After a number of sales scams in the 1980s, aluminum siding salesmen were viewed as lower than used car salesmen. Aluminum awnings in the 1980s were being banned by homeowner associations.

The decline of aluminum's image appeared in many unusual places, such as coinage. Prior to the Hall-Heroult process, the United States issued an experimental proof set of aluminum coins that was an equal of gold and silver coinage. Germany had minted aluminum bronze coins as early as 1918, when the metal still was considered valuable. Right before and after World War II, aluminum coins were minted in much of Europe as a sign of the nation's level of technology. Norway continued the minting of aluminum bronze coins into the 1950s. The People's Republic of China began issuing aluminum coins in December 1957, in denominations of 1, 2, and 5 fens. These Chinese coins became a symbol of cheapness. By the 1960s, consumers were rejecting aluminum in coins around the world. In 1974, the United States Congress was ready to change to an aluminum penny. The potential savings were huge with a 96 percent aluminum penny replacing the copper ones. The public resisted, and Congress voted the change down. While there was obvious resistance from the copper industry, the general public reportedly didn't like the feel and the perception that it was of a lesser value. Ironically, a few of these aluminum sample pennies seemed to have been pocketed by some congressmen and command a king's ransom today.

By the late 1970s, aluminum had an image problem. It had been the metal of the future, of the metal artists, and a symbol of technology and progress in prior decades. Aluminum had put a man on the moon, but aluminum in the 1960s had become more utilitarian and commonplace. Aluminum had become part of life for the middle class. Becoming commonplace took away from aluminum's allure. It was everywhere by the late 1960s. Suburban kitchens and homes had aluminum pots, pans, and aluminum foil, packaged aluminum popcorn poppers (Jiffy Pop), TV dinners packaged in aluminum, aluminum awnings, Eskimo Pies and Klondikes wrapped in foil/paper, cans, and aluminum folding chairs. Grandparents of the 1960s lamented the replacement of real Christmas trees with aluminum ones, but the aluminum tree proved to be just a fad.

Aluminum Christmas trees were first commercially manufactured in 1955. The trees were manufactured by Modern Coatings, Inc., of Chicago. Between 1959 and 1969, the bulk of aluminum Christmas trees were produced by the Aluminum Specialty Company in Manitowoc, Wisconsin. The trees retailed for $25. The trendy aluminum tree was highlighted in the TV hit, *A Charlie Brown Christmas*, in 1965, which would also be the peak year for aluminum trees. In that episode Lucy told Charlie Brown their play needed a great big shiny aluminum Christmas tree! Maybe painted pink!" By the 1970s, aluminum trees were history. The pink aluminum Christmas tree, along with the plastic pink flamingo, became the icon of the lower middle class. Once the metal of kings, aluminum was now the metal of the common man. In the 1990s, the TV show *Seinfeld* took a final shot at the aluminum tree with its aluminum Festivus pole and holiday.

By the 1970s, the growth of trailer parks further associated aluminum with the lower class. The success in the 1960s of aluminum resulted in over 25 percent of the residential single-family homes being mobile homes in the 1970s. The trailer could be mass produced on assembly lines, driving the price down. Fuller's dream was being realized, but aluminum's image was on the decline. Mobile homes became wider and longer, moving them from semi-permanent to permanent structures. Trailer parks continued to multiply into the 1980s, but then the public turned on trailers. Trailer parks tended to reduce house prices in nearby residential areas.

The faded aluminum trailer with torn aluminum folding chairs in front caused new zoning laws to be passed. Aluminum trailer parks became the icon for suburban ghettoes. "Trailer trash" would become a term for poor whites. Aluminum had become a low-price commodity. There was a price drop in the 1960s as President Johnson opened up the strategic stockpile to fight inflation. The Johnson administration dumped over 200,000 tons of aluminum on the market out of its war stockpile of 1.4 million tons. Production of aluminum trailers and aluminum chairs boomed as prices rapidly decreased, and the reputation of the once-valued metal bottomed.

Another problem for aluminum's image was consumer issues. Fires from aluminum wiring would make the daily headlines. Trailer homes were using aluminum wire in the late '60s and early '70s. The use of aluminum as conducting wire was not new, but the problems of mass commercialization were. Aluminum-based electrical wiring was first used for utility distribution applications in the early 1900s. Use of aluminum wiring grew rapidly after World War II and has increasingly replaced copper as the conductor of choice in utility grids. The metal has significant cost and weight advantages over copper, plus a superior conductivity-to-weight ratio, making it the preferred material for electricity transmission and distribution uses. Its use in trailers spread to general housing. Aluminum wire was used to wire entire houses for a short time from the late 1960s to the late 1970s during a period of high copper prices. Wiring devices at the time were not designed with the particular properties of aluminum wire in mind, and there were problems with the properties of the wire itself. Aluminum wire connections also required different techniques from those employed with copper; this was often overlooked initially, which became the root of the problem.

Over time, many of these terminal connections to aluminum wire began to fail due to improper installation, poor connection techniques, and dissimilar metals having different resistances and different coefficients of thermal expansion in the system. These connection failures generated heat under electrical load and caused overheated connections. A number of fires started to be attributed to aluminum wiring. On April 28, 1974, two persons died in a home fire in Hampton Bays, New York. Fire officials determined that the fire was caused by a faulty aluminum wire connection at an outlet. Publicity caused consumer panic and ended the use of aluminum wiring for decades. As a metal, aluminum was now tarnished, and again it was associated with cheaper uses. ALCOA would resolve the problem, but it is only in recent years that aluminum wiring is again being used in housing due to consumer resistance and fears. A major part of the problem was the alloy being used. After extensive testing, a series of aluminum alloys suitable for use with electricity were developed in 1974 by ALCOA. These alloys were designated as the 8000 series. In addition, the Aluminum Association and engineering societies set a number of standards for installation. While the technical issues were solved, aluminum's image would never be the same.

By the mid–1970s, the image of aluminum had clearly changed. Author Mini Sheller personally recalled the change in how aluminum was viewed in 1970s: "I also remember the beginning of the change in our feelings about aluminum. As the space age passed and the oil crisis set in, chairs began to fall apart and the aluminum siding developed stains. We crushed cans under our feet to turn our sneakers into funny noisemakers and made craft projects from pull tabs that still detached from the lid in those days. This flimsy metal began to seem worthless, recycling campaigns seemed to increase its association with trash rather than with the chic modern design of previous decades."[1] The image would slow the application in the new technology of the housing industry.

This poor image of aluminum and trailer homes would create a resistance to the idea of the modular home. A reporter for the *New York Times* noted the problem in 1986: "People think of trailer when they think of modular … a sort of stigma as something not permanent."[2] In the '50s, '60s, and '70s, most American homes were built on site from "stick" up. The process included many subcontractors such as masons, carpenters, and plumbers. Factories could more easily manufacture modular units or parts to be assembled on site. The concept of modular had been developed by Kaiser in building Liberty Ships for World War II, and it offered a huge market for aluminum. But consumer resistance was high against such an approach. The *New York Times* summarized the continuing image in 2006: "Still, modular and prefabricated housing struggled with the image of the mobile home and the trailer and of assumed inferior construction; though modular housing companies could build mansions and large single family split levels, the same factories also continued to produce mobile homes, thus making conflation nearly inevitable."[3] While sales had slowly increased, the resistance still remained on the high end of the market.

One victory of the 1970s was the aluminum bat. The first metal baseball bats were made in the mid–1920s, but performed poorly. They would often bend and dent if solid contact was made. Aluminum bats became good enough for mass distribution in 1970. Aluminum alloy bats had the same properties of energy transfer that had made aluminum popular in tennis rackets. Early bats were made of the historic 6061 alloy, which lacked dent resistance in long-term use. Aluminum companies entered a development program to meet market demands. In 1996, the top-of-the-line softball bats offered by the bat companies were made out of an alloy designated Cu31 (a special ALCOA designation). This extremely popular aluminum alloy of zinc, magnesium and copper alloy is more commonly known as 7050. The trademarked designation Cu31 means that not only is the alloy 7050 but that it was supplied by ALCOA. This alloy was a variation of ALCOA alloys in the 1920s for aircraft propellers and the top-secret aircraft alloy of Japan's Sumitomo Metals in the 1930s.

Worth Sports Company supplied the first aluminum bat to Little League baseball games. Later on, another bat company, Easton, manufactured their first metal bat, which was used by the United States Olympic team as they went on to claim the gold medal. In the 1970s, high schools and colleges started to experiment with aluminum bats. The revolution continued, with Louisville Slugger quickly joining after the NCAA legalized the use of aluminum baseball bats in 1974. With the use of aluminum bats, team batting averages went up about twenty points and home-run production also doubled. The performance is one reason that the major leagues do not allow aluminum, since historical records need to be protected.

The 1990s brought renewed interest in aluminum bicycles. Aluminum had, for decades, held promise in bicycle production because of its lightness, but it never lived up to its expectations. In the 1930s, Duralumin bicycles became popular but required mechanical joints. Welded steel frames had a major cost advantage. Aluminum frames had started to be considered with welding improvements such as TIG (Tungsten Inert Gas Welding) and special heat treatments, to be applied after welding, developed during World War II. Amazingly, it would be Reynolds Metals that would put steel over aluminum for decades in cycling. Reynolds Metals developed a special steel alloy for frames in the 1950s. In 1958 Charly Gaul of Luxembourg took a victory in the Tour de France with a frame built from Reynolds 531. This would mark the beginning of what would soon become a near total domination of the Tour de France by Reynolds steel frames. Between 1958 and 1991, Reynolds tubing was used in the bikes of 26 Tour de France winners. The potential of the weight advantage of aluminum remained a source for innovative research. An aluminum frame would win in the 1990s, but it never took off with the public. Aluminum bikes for consumers were plagued by cost and manufacturing problems. In addition, new high strength, low alloy steels became even more competitive in lightness and cost.

The major advance was made in 1975, when Gary Klein displayed his welded and heat-treated aluminum frames at the International Bike show. Italian bicycle company Alan and French were producing their lugged aluminum frames around the same time; and, in 1983, Cannondale launched their "Aluminum for the Masses." These models used thicker walls to achieve durability and stiffness comparable to that of steel. Still, welded joints offered stress points for fatigue breakage. Cannondale's solution was a TIG welded 6061 alloy, which was heat treated after the frame was welded.

Other competitors in aluminum bikes eliminated the welding for mechanical joints. The framesets were made of small-diameter 5086 aluminum alloy tubing that was slip-fit onto aluminum sockets and then glued into place using a heat-activated type of epoxy. American Manufacturing, out of St. Cloud, Minnesota, put aluminum tubing on the average bike with sparkly mid-sized tubes in 1991. While aluminum bikes offered weight savings, they still were expensive at twice the price of steel. Cannondale's advanced aluminum design avoided fatigue problems often noted in other bikes.

Aluminum in the 1990s became extremely popular with mountain bikers. It had favorable strength-to-weight ratio and a lower cost compared to titanium, steel, and composites used for bicycles. However, when compared to these other materials, aluminum is more susceptible to fatigue failure at lower cycle count. Fatigue causes catastrophic breaks. Fatigue failures that occur during typical usage of mountain bikes had devastating effects for bicycle manufacturers, resulting in expensive recalls, legal liabilities, and loss in product image. Facts suggest it was more of a scare than reality. Aluminum, however, started its decline with these published failures as composites and titanium made inroads. Cannondale still makes an aluminum frame called the CAAD 10. Cannondale's CAAD 10 uses the new alloy 6069, which some believe will once again compete with carbon fiber in weight and price. In the 2010s, aluminum is once again making inroads into mid-price models between steel and carbon fiber. The 6000 series aluminum provides a better ride than earlier alloys.

The 1990s were turbulent times for the aluminum industry. In 1999, Alcan made a failed attempt to make a three-way merger between it and Algroup (Alusuisse Lonza Group) of Switzerland and Pechiney of France. The proposed merger was blocked by the European Commission due to fears of anti-competition. After the deal fell through, Alcan acquired Algroup in 2000. Then in 2003, Alcan acquired Pechiney, completing the original three-way merger plan of 1999.

On the product front, the search for the Holy Grail of aluminum car bodies continued in the 1990s. In the 1980s, the progress toward aluminum car bodies was slowed by aggressive product development by the steelmakers. The development of high-strength, low-alloy steels allowed for thinner gauge steel. In addition, the electro-galvanizing process all but stopped the corrosion of steel. The overall advantages of aluminum included a major benefit of reduced emissions (via weight reduction). The environmental movement continued to keep an eye on the situation, and in the 1990s, aluminum surfaced again as a competitor to steel. This was particularly true in Europe, where fuel prices remain much higher than in America. The aluminum car has been trying to make a comeback since the Pierce-Arrow of the 1920s. However, it was mostly piecemeal until the 1990s.

In 1992, Ford began experimenting with aluminum with its Aluminum Intensive Vehicle program, which led to the development of an all-aluminum-body Taurus. Ford actually manufactured 100 of these cars, but aluminum and oil prices again restricted going to full production. In Europe and America, Ford developed the aluminum concept car, the Ford Contour. The body, created from a chemically bonded modular aluminum structure, was invented by Ford and Reynolds Aluminum. The body panels form composite structures incorporating polyurethane and plastics with metal framing. The name "Contour" was appropriate, as the body does not possess a single flat line or panel.

The aluminum Jaguar had employed Alcan's Aluminum Vehicle Technology (AVT) structural bonding system on the XJ220 sports car, produced from 1992 to 1994. The Audi A8 made in 1992 is notable for being the first mass-market car with an aluminum chassis; all A8 models utilized this construction method co-developed with ALCOA. The chassis became known as the space frame because of the alloys, which had been developed for the space program. It was also manufactured with an all-aluminum body. The Audi A8 used seven different aluminum alloys to achieve a car 40 percent lighter than an equivalent one in steel. The aluminum super frame used the newly developed aluminum-magnesium-silicon extension alloys like 6082. This extension had, at their root, the old Duralumin composition with heavier additions of silicon and magnesium. But the real technology was in how these extruded alloys could be heat-treated to high strength levels. It took nearly 10 years for Audi to reach 150,000 aluminum cars.

In Europe, the aluminum frame and body made further inroads in the high-price end of the sports car market. Ferrari began using 6000 series aluminum with the company building all its current production models—the 458 Italia, 458 Spider, 599 GTB, California, and the F—from aluminum. Ferrari used an epoxy bonding system with aluminum, which worked better than carbon fiber, at volumes of 30 cars per day.

Aluminum also made inroads into rapid transit trains as Japan introduced environmentally friendly, recycled aluminum cars in 1994. Of course, aluminum was not new to rapid transit; Toronto introduced aluminum subway cars in 1954. In the 1960s, Chicago's El system used aluminum in the frames and car exteriors. As Japan and Europe aggressively started to pursue aluminum alloy development in the 1990s, the very nature of the industry was changing. Old names like ALCOA, Reynolds, and Alcan remained, but now Russian and Chinese companies became competitive.

CHAPTER 21

The World Aluminum Wars

THE 1990s WOULD BRING A VERY DIFFERENT world aluminum industry. Still, the old factors of cheap energy and abundant bauxite deposits determined who controlled the world market. New leaders such as Russia and China would emerge. The 1970s, with the change of aluminum to a commodity, brought with it changes to the aluminum companies. In the 1970s, the Big Six of ALCOA, Alcan, Reynolds, Kaiser, Pechiney, and Alusuisse controlled 60 percent of bauxite mines, 80 percent of the alumina production, and 73 percent of primary aluminum production. The global primary aluminum industry of the early 1970s was highly concentrated and vertically integrated. The 1980s would bring market challenges for the international as well for the American aluminum industry. The progressive lowering of tariffs throughout the world under the General Agreement on Tariffs and Trade had been a blessing in the 1960s and 1970s for the American aluminum industry; but the blessing turned curse finally came in the 1980s.

In the 1980s, the new internationalized market opened the door to cheap finished aluminum products in the United States. The American aluminum market was still the world's largest, so loss of market to foreign companies became a real problem. Inflation and economic conditions were problematic during the 1980s. Again the American aluminum companies fared better than most. One reason was the inability of the Japanese aluminum to challenge American producers as they did in autos and steel. The high power costs of aluminum had effectively held Japanese out. The Japanese lacked hydroelectric resources and depended on coal and nuclear. By 1980, Japanese smelters were particularly vulnerable, and Japan closed down 75 percent of its smelting industry during the decade. The Japanese advantage in labor meant little in the low labor input of aluminum production. The Japanese looked for partners. Canadian Alcan, with its extremely low energy costs, did take market share and also became a source of low-cost ingot for Japanese and European fabricators. Canadian energy was a mere 15 percent of the cost of energy in the United States and 5 percent of the world's average cost. Those hydroelectric plants of Alcan were finally paying off. And as bad as things were, after the war of developing hydroelectric power, ALCOA's strategy put them in a very competitive position with Kaiser and Reynolds as well as the smaller domestic companies. Aluminum had become a commodity and its image would start to decline in the world, but the root of success in the aluminum industry continued to be process and product innovation.

The 1990s would also show that aluminum still had the ability to create millionaires and billionaires. Probably the most colorful part of aluminum's history was the famous Russian aluminum wars of the 1990s. In the aluminum war, rivals fought for control of Russian smelters in the 1990s. Aluminum plants remained cash cows and sources of huge wealth for owners. These days are long gone and the wars are over—but the emergence of RUSAL as Russia's only aluminum company was rooted in these wars. The aluminum wars were played out in a period where Russia was struggling to come to terms with the transition from the old Soviet-style economy to a new free market.

Growth of the Russian aluminum industry under the Soviets was a great achievement. Russia had started to develop its industry in the 1930s. In the 1950s, new aluminum smelters were built in Kandalaksha, Nadvoitsy, and Volgograd, where cheap hydroelectric power was available. In the 1960s and 1970s, smelters in Irkutsk, Krasnoyarsk, and Bratsk were constructed in close proximity to the largest hydropower plants in Siberia. Cheap electricity from hydropower in Siberia was and is at the heart of the Russian aluminum industry. That power cannot be transmitted via high-tension cables because it is too remote; instead, it is used in local aluminum smelters. This cheap power made Russian aluminum capable of achieving huge profits. By the early 1980s, Russia was the world's second largest producer of aluminum after the U.S. With the fall of the Soviet Union came a new type of business, part capitalism, part crime syndicate, and part political corruption. It was a much different type of capitalism than that of the historic origins of aluminum. This was more a corrupt business that lacked government regulation. In reality, aluminum had been moved from the black market under communism to the white market of a corrupt socialism.

In 1993, the Russian government launched the privatization of the aluminum industry. The change from communism to an open market economy proved awkward at best. International traders who obtained access to Russia's largest aluminum smelters during the privatization were not interested in developing the sector and did not invest in production, opting for immediate profits instead. The Russian investors proved corrupt and opportunistic. Old-line socialists still had most of the old money available for investment. This period in Russian business history is now known as the "aluminum wars."

Russian businessman Oleg Deripaska started to acquire and improve massive Russian smelters, building a vast aluminum empire. Deripaska had deep ties with the old and new money. Deripaska was a member of the ruling Boris Yeltsin family through marriage. He was also part of a wealthy and secretive class of Russian businessmen known as "oligarchs." Deripaska built an aluminum empire from factories and mines that had been privatized individually, by forcing out weaker owners. In 1997, as part of a general restructuring of the companies controlled by Mr. Deripaska, Sibirsky Aluminum was established to manage aluminum and alumina assets acquired by companies related to Mr. Deripaska. Deripaska continued in a battle among Russian investors to take control of the industry. Deripaska had made a fortune worth billions by his investing in aluminum. He argued that his huge profits in aluminum attracted gangsters and the need for protection money for managers and executives as a new corrupt version of capitalism morphed out of communism.

The main struggle was between two rich factions—Oleg Deripaska and Lev Chernoi, and on the other side Roman Abramovich and Iskander Makhmudov. Abramovich and Makhmudov had made their money in oil. The two sides hired gangsters to help in their fight to rule the aluminum industry. The fight was a type of gangster capitalism which included tax evasion, bribery, political payoffs, and violence. It was clearly more reflective of the 1930s battle over the liquor industry by the crime syndicates. Managers, bankers, and journalists were killed, and the deaths were over a hundred in this takeover struggle. Assassinations were threatened of the main players, and these capitalists were heavily armed. The *New York Times* described some of the action: "Within a year of his appointment in 1994, for example, Mr. Deripaska's financial director survived an attempt on his life in Moscow, apparently by a hired killer. The director of a RUSAL smelter nearby in Krasnoyarsk quit after being beaten nearly to death in the entryway to his apartment. Later, another factory director was accused of ordering the assassination of a Siberian governor. So lawless was Siberia that one aluminum factory changed hands literally with a keystroke, when one large shareholder was deleted from a database of owners and had little recourse in the powerless courts."[1]

The Russian politicians turned their heads, probably because they were getting plenty of money spread around. Vladimir Putin, who was sensitive to political interference from ambitious oligarchs, has condemned more than one to prison or exile. Putin forced a Deripaska/Abramovich consortium as a compromise, and the metal murders ended. The consortium would be strengthened by the government and become RUSAL. The Russian aluminum industry became an example of national socialism more like that found in Germany of the 1930s. Its major strategy has been to purchase bauxite resources, which the Russians had lacked. RUSAL has established its presence on five continents, including in Guinea, Australia, Guyana, China, and Nigeria. RUSAL began like the Russian oil industry, an extension of Russia's international geopolitical power.

RUSAL today is a true heavyweight in the international aluminum industry. RUSAL is the global leader in the aluminum industry in production. In Siberia it operates five huge aluminum smelters and two massive refineries— but its influence now goes a lot further and it's now a global player. RUSAL recently overtook ALCOA and Alcan as the largest aluminum producer in the world after merging with its nearest rival SUAL (Siberian-Urals Aluminium Company). The new RUSAL is the world's largest aluminum and alumina producer, with 12 percent of the world's aluminum market, 15 percent of the global production, operations in 19 countries, and around 100,000 employees. RUSAL, in the last few years, has struggled with the Aluminum Company of China for the top position.

Things had once again trended differently with the turn of the century. The BRIC countries (Brazil, Russia, India, and China) were producing a third of global production in primary aluminum in the year 2000, with Russia accounting for 13 percent of the global total. In 2012, the BRIC contribution had surged to 56.6 percent of global primary production, with China being, by far, the largest producer. During the same period, the cumulative share of Japan, Western Europe, and North America was sliced by half from 40 percent to less than 21 percent. As for primary consumption, the BRIC share surged to 48 percent, while the share of major industrialized consuming countries went in the opposite direction, from about 60 percent in 2000 to below 33 percent.

Primary aluminum production has been moving to China, which accounts for 40 percent, not only because of its advantage of cheap and abundant energy, but also because of government induced sources of competitiveness related to subsidies, exchange rates, and trade policies. The predominant primary producer in 2012 was the Aluminum Corporation of China Ltd. (Chalco), while three other primary aluminum producers, Shandong Huaxin Aluminum Company, Shanxi Guanlv Holding Co. Ltd., and Qingtongxia Aluminum Group Co., have significant production capacity as well. American companies are now paying the long-range price of the global trade agreements of the 1970s. China's position is not so surprising since it now accounts for 40 percent of the consumption. China is well placed to expand in the global markets, since it has cheap energy and extrusion costs are estimated to be 60 percent of those of extrusion in the United States. In the 1970s, more than 60 percent of global consumption of primary aluminum was taking place in six industrialized countries, with the United States leading the pack at 36 percent, followed by Japan, Germany, France, Italy, and the United Kingdom. In 2012, the combined share of these same industrialized countries was barely exceeding 25 percent. RUSAL was the largest aluminum producer, followed by ALCOA, Alcan, and Chalco in 2014. Today RUSAL and Chalco battle back and forth for the top producer position, but China, as a whole, is the largest producer. China controls the market and price to a large degree. American companies are now held back by internal antitrust laws, while having to compete with huge Russian and Chinese government-backed monopolies.

China and Russia are far from alone with their aluminum dreams. Aluminum still remains a symbol of and a real basis for industrial power. India, much like Japan, is a country in search of primary aluminum sources. It lacks the energy to set up primary smelting, since it is basically a coal-based country. India is looking to team up with Iran to build a smelter. Iran will be using a mix of sources for power generation, including oil products and natural gas, even some hydroelectric. The world is changing as countries look to become aluminum independent. Countries like China and Russia are creating a huge surplus.

There had been some changes in the traditional companies as well. In 2008, Alcan Inc. was amalgamated with Rio Tinto Canada Holding Incorporated, following the latter's acquisition of a majority of the share capital in Alcan. Following the acquisition, which was carried out by way of amalgamation, Rio Tinto Canada Holding was renamed Rio Tinto Alcan Inc. Anglo-Australian Rio Tinto is the world's second biggest mining company, behind BHP Billiton, producing iron ore, copper, diamonds, gold, coal, and uranium.

The geographic distribution of bauxite mining has shifted significantly since the 1970s. It is a far different world from that of Charles Hall. Bauxite mining in Australia increased its share of global output from 20 percent in the 1970s to 32 percent in 2012. Amazingly, Jamaica, Surinam, and Russia are no longer on the list of the major producers, having been replaced by Brazil at 15 percent, China at 14 percent and Indonesia at 11 percent. The combined market share of the four largest bauxite producers is now over 70 percent. In the last few decades, however, the focus on health, environmental issues, and the plight of the small nations and miners has grown.

Most large bauxite mines are in places like the outback of Australia and have survived detailed scrutiny. Mining of bauxite is not labor intensive as with most metal ores. The Caribbean, however, is one of the mining regions that intersect with populated areas, rich tourists, international politics, and the international press. Jamaica is one of the few major bauxite sources where the mines are in populated areas. The environmental and ancillary health problems are readily visible. Jamaica received its independence in 1964, so this can no longer be a matter of colonial capitalism after a socialist revolution took place. In 1974, the government became a joint venture partner with ALCOA under nationalization. Jamaica has turned a blind eye because, after tourism, bauxite mining represents its largest source of income.

In 2004, the *Los Angeles Times* summed up the situation:

> Residents of the bauxite—alumina sites, mostly in these undulating southwestern hills around Mandeville, have complained throughout the industry's half a century of operations here that their ailments stem from exposure. Health studies elsewhere have linked bauxite to hypertension and alumina dust to asthma and sinusitis. Jamaican authorities dismiss the complaints of illness. Officials reject requests for compensation, medical treatment, or corrective measures on the grounds that there is no statistical proof of causation from the processing of bauxite into alumina, the key element for making aluminum. Complaints from thousands of Jamaicans about asthma, sinusitis, and children with birth defects have prompted a militant minority to challenge what it describes as the Caribbean nation's see-no-evil policy.[2]

This issue will continue to be a problem, but the massive bauxite deposits in Australia put pressure on everybody to see no evil.

Another criticism of bauxite mining, leveled by international socialists, has been the siphoning off of wealth to the more developed countries. It, of course, is an old problem; socialists had attacked America in the 1930s for their South American mining operations, which supplied North American manufacturers. Socialists maintain this is a bigger problem in Jamaica. This has become a hotly debated issue in academia. In defense of the years prior to nationalization, ALCOA, Kaiser, and Reynolds moved their alumina processing plants to Jamaica. Jamaica can no longer be considered a victim of colonial capitalism; the socialist government owns part of the problem now. Considering the destructive deindustrialization of American industry, it is hard to argue that true wealth has been transferred. Jamaica is probably an example that corruption knows no political boundaries; it can affect both socialism and capitalism. While not overlooking the problems of poor Jamaicans, the problem, not the politics, should be addressed.

In January 2014, Indonesia imposed its ban on the export of unprocessed minerals such as bauxite. The ban was to generate greater value for the country and its citizens by forcing operators to build processing plants and export value-added product, not raw materials. Indonesia has only lost business as a result. The reason is the vast geographic distribution of bauxite. The main target was Chinese aluminum, but instead of investing in refining plants in Indonesia, the Chinese stepped up Australian bauxite purchases. Prior to the ban, most bauxite for China came from Indonesia. Indonesia needs to attract Japanese investors, who are still looking to build an aluminum industry.

Aluminum smelting has also seen its share of environmental problems. The *Washington Post* reported on problems in Mongolia:

> A few years after the smelter opened, herders in the area said that their sheep began falling sick, with jaws so painful that they could not eat. Soon, thousands of their animals had died. When they complained, the government simply arrested five of their leaders and forced the others to resettle in the nearby city of Holingol, demolishing their original homes. The vast, wind-swept grasslands of Inner Mongolia have been home to nomadic pastoralists for thousands of years, but the rich resources that lie under these rolling prairies have proved a curse to the people who have long called this land their home. A boom in mining and mineral industries has polluted the grasslands, marginalized herders, and pushed them from their homes. Now, a fall in coal and gas prices could spell more pressure on government spending and more misery for herders.[3]

The herders had gained nothing from the building of the smelter. After weeks of violent protests, the Chinese government decided to shut down the smelter.

Problems of this type will continue as the world's appetite for aluminum and industrial manufacturing increases. We may once again look at the use of paternal capitalism (as imperfect as it was), at least as a start. There is no doubt that globalization has brought coldness to relations with local communities, and no political or economic system has all the answers.

CHAPTER 22

The Future, Science Fiction and New Uses

ALTHOUGH THE HALL-HEROULT PROCESS supplied the commercial breakthrough to open up the aluminum age, its application is still limited by economics and availability of natural resources. And, while the energy necessary to smelt aluminum has decreased 70 percent in the last century due to more efficient processes, electricity still remains a key variable cost in aluminum production. The huge amount of electrical power to manufacture aluminum and its related costs has limited it from making inroads into traditional steel markets and kept whole nations out of the aluminum industry. The overall consumer cost benefits, such as weight savings and reduced fuel consumption, put aluminum on the edge of competing with steel in auto sheet production.

Aluminum is looking once again to the auto industry. Ford has renewed the interest in the aluminum car with its announced use in its F-150 truck in 2014. The cost premium to switch to aluminum over steel is about $500 per truck. Ford will shed more than 700 pounds from its latest F-150, about 15 percent of the vehicle's body weight. That switch has the potential to boost fuel economy up to 20 percent, depending on the model. Not only does that save consumers a bundle on gas, but it's also better on the environment, as improved fuel economy cuts down on greenhouse gas emissions. Aluminum has become the poster-child metal for green consumers. It is still creating new markets and energizing older markets.

Jaguar has again become a pioneer in aluminum-bodied cars, bringing a lightweight production car, the X350-series XJ, to market in 2003. The adhesive bonding system has been more successful than welding. Almost 90 percent of the XJ's body is made from various grades of aluminum, the balance being steel and magnesium. The C-X17 sports crossover concept introduces the next generation of Jaguar's advanced aluminum architecture. Sales have been slow, however.

Tesla Motors introduced the Tesla Model S electric vehicle in 2013, an all-aluminum premium sedan that accelerates from 0 to 60 miles per hour in 5.6 seconds, with zero tailpipe emissions. The body of Model S is a state-of-the-art, aluminum-intensive design. Weight-saving benefits make aluminum a natural choice for this electric car. Stamped sheets of aluminum, extrusions, and castings are expertly joined for rigidity and strength using adhesives and welding. A rigid and strong structure not only protects passengers but also contributes to overall maneuverability behind the wheel.

Aluminum bodies have not lived up to sales predictions in most cases. Some problems remain at the raw material level as aluminum is three times as costly as steel is. In the stamping or conversion process, aluminum is twice as costly as steel. In assembly, aluminum is 20 to 30 percent more expensive. In total, steel uses are half the costs of aluminum. And of course, there are capital costs to completely convert stamping operations to aluminum. The main advantage of aluminum is significant weight reduction, which translates into reduced fuel costs and reduces emissions. The green movement remains a major factor in aluminum's future.

The problems of aluminum repair have been a problem in gaining sales. Companies like Tesla are offering training to repair shops using the Fronius cold metal transfer welder used in building the car. The problem for repair shops as well as manufacturers is it requires a major investment in tooling and equipment. Chrysler has an aluminum design for its Jeeps, but energy prices are key for it to become reality. For Chrysler, it would require a new factory for aluminum body production. The dramatic drop in oil prices in 2015 halted a decision by Chrysler to opt for aluminum bodies. For the consumer, the higher price of aluminum has to be offset by fuel savings, making aluminum cars sensitive to the price of oil.

General Motors has launched an advertising campaign to take on Ford's use of aluminum in trucks. "Chevrolet is escalating the always-hot pick up wars this week with a social media campaign that takes a swipe at Ford's aluminum-body F-150. In a commercial for online and possible TV play, focus group members are asked which cage they would use for refuge when a grizzly bear lumbers into the room: one made of aluminum or another made of high-strength steel. Most scurry wide-eyed to the steel cage."[1] It deals with the age-old perception that aluminum is not as strong as steel. That perception is rooted in the common consumer's experience with aluminum cans, siding, gutters, and wire. Many are calling this struggle a new type of aluminum war.

In 2012 NASA started research on 3D printing for aluminum parts. Divergent Micro Factories, a new startup company, built the first 3D-printed aluminum car chassis in June of 2015. The body is fiber composite. Instead of 3D printing an entire vehicle, they 3D print aluminum "nodes" which act as the joining mechanism for fiber tubing. Straight fiber tubes can be joined to make any shape. The 3D printing allows for elaborate and unusual-shaped nodes, which are then joined together by off-the-shelf carbon fiber tubing. The aluminum nodes supply the needed strength and flexibility at the joints, where all-fiber tubing could not. Once the nodes are printed, the chassis of a car can be completely assembled in a matter of minutes by semiskilled workers. The process of constructing the chassis is one which requires much less capital and other resources. The body is then added. The 3D printing of aluminum fiber chassis could allow for micro auto factories in the future.

Aluminum auto parts continue to increase in popularity every year. A spinoff NASA aluminum alloy developed in 2003 is making inroads in many high-temperature applications such as auto pistons. A high-strength aluminum-silicon alloy was created at NASA's Marshall Space Flight Center (MSFC) in Huntsville, Alabama. The alloy contains 18 percent silicon plus minor amounts of strontium. MSFC-398 is a newer version of NASA aluminum alloy 398 that has 2 to 3 times the strength of cast aluminum pistons at 700 degrees. Honda started using this alloy in their pistons in their E-Tec motors. It is also finding application in high-temperature two-cycle engines. At 600° Fahrenheit, it exhibits 3 to 4 times the strength of conventional cast aluminum alloys. The space agency projects that the new metal can be produced at about one dollar per pound. This amazing alloy holds its dimensional stability at higher temperatures, which has made it a material for high-temperate exhaust fans in many applications. NASA scientists feel that 398 could be the alloy for the future Mars shuttle *Gen1 Enterprise* hull.

NASA has played a critical role in focusing national research on aluminum. NASA is our peaceful version of Wilhelm II's German scientific research center at Neubabelsberg in 1902. It was at Neubabelsberg that the revolutionary alloy of Duralumin was developed. In the 1920s, England and France also developed a government and private company to develop war alloys. Today, NASA serves to build these necessary partnerships between government and industry. Aluminum research at NASA has served as a model for cooperative capitalism.

New alloys continue to increase the overall usage of aluminum in cars. The increase is about 5 pounds a year and averaging 345 pounds per car in 2015, mostly into engine blocks, wheels, suspension components, and hoods. Almost 70 percent of all cars sold in North America now have aluminum engine blocks, and over 22 percent feature aluminum hoods. Maybe the biggest future for aluminum in cars is the aluminum-air battery seen in the Tesla Model S electric car tests of recent years. Aluminum-air batteries produce electricity from the reaction of oxygen in the air with aluminum. They have one of the highest energy densities of all batteries, but they are not widely used because of problems with high anode cost and byproduct removal when using traditional electrolytes; this has restricted their use to mainly military applications. However, an electric vehicle with aluminum batteries has the potential for up to eight times the range of a lithium-ion battery with a significantly lower total weight. Aluminum-air batteries are primary cells and are not rechargeable. Once the aluminum anode is consumed by its reaction with atmospheric oxygen, the battery will no longer produce electricity. However, it is possible to mechanically recharge the battery with new aluminum anodes made from recycling the hydrated aluminum oxide produced in the battery. Such recycling would be essential if aluminum-air batteries are to be widely adopted. Stanford University has recently announced the development of an aluminum-ion battery that is rechargeable.

Aluminum applications in aircraft are far from over, too, even though composites have made inroads. The Airbus A380, one of the largest passenger airliners in the world, contains 10 times the amount of aluminum used in the Airbus A320. And Boeing's 787 Dreamliner, which is often described as a composite aircraft, contains 20 percent aluminum (by weight) which includes aluminum 7085, a relatively new aluminum alloy. The application of high-strength, lightweight aluminum-lithium alloys could reduce weight in large airliners by up to 20 percent. Airbus A350 uses a great deal of aluminum-lithium for the wings and fuselage. Boeing is following Airbus. ALCOA has a new $90 million advanced manufacturing facility that produces more than 44 million pounds of aluminum-lithium for the production of airplanes.

Research into aluminum-lithium (Al-Li alloys) began in the U.S. and Germany in the early 1920s. Super aluminum-lithium alloys were even studied in Wilm's German Research Center. During World War I, several patents were given during this time; however, the alloys had such poor performance properties, they were never used commercially. No major developments were made with the alloys until the late 1950s, when the ALCOA Corporation patented the 2020 alloy. This alloy can be considered as the original Wilm super alloy Duralumin with lithium added. Alloy 2020 contained 4.5 percent copper, 1.2 percent lithium, .5 percent manganese, .2 percent cadmium, with the remainder aluminum. The Navy initially used some 2020 in experimental aircraft. The 2020 alloy and the previous patented alloys are called the first-generation aluminum-lithium (Al-Li) alloys.

In the 1970s, the reduced weight from adding lithium (Li) drew interest again in saving fuel in aircraft. The 8090 alloy, the most successful of the second generation, was used in EH101 helicopters in Europe. However, they typically failed to meet full aircraft standards. In the 1990s, a third generation of Al-Li alloys was developed. The first third-generation alloy, Weldalite 2195, was used for a U.S. space shuttle external fuel tank. This new alloy, besides lithium, had minor amounts of silver and magnesium. The reduced density and high strength were the primary reasons for its use. These low-density alloys are attractive to the aerospace industry since structural weight reduction is a very efficient means of improving aircraft performance. ALCOA has improved on Weldalite 2195 with alloys 2090 and 8090. Alcan more recently has developed a patented Al-Li alloy. The shuttle's fuel tanks used these alloys and saved several tons in tank weight. The future seems to be bright for these amazing alloys.

In more standard transportation such as railroads, aluminum tonnage is increasing due to these improved wear-resistant aluminum alloys. Deliveries of rail cars have averaged some 60,000 units in the last 10 years, projecting a 50 percent growth in coal car deliveries over the next few years. The booming coal shipments to China are driving the use of aluminum freight rail cars. The lighter aluminum rail car bodies, about two-thirds the weight of the comparable steel body, enable a greater payload. The higher payload capacity repays the higher initial cost of aluminum in less than two years, and the resistance of aluminum to corrosion by the high-sulfur coal ensures long durability for these coal cars over traditional steel. About 30 percent of coal cars are now all aluminum.

Aluminum once again is getting consideration in boat building. The Japanese Coast Guard has used aluminum boats from the 1950s. Today the Japanese National Guard has several all-aluminum patrol boats such as the *Tsurugi*, which is a 164-foot-long patrol boat capable of a maximum cruising speed of over 40 knots. Australia is currently building a series of all-aluminum Armidale class patrol boats for the Royal Australian Navy. Australia has also been a leader in aluminum ferryboats. Australian builders are praising aluminum deep V hull designs for the future. According to marine design engineer Michael Kasten of Kasten Marine Design, Port Townsend, Washington, "An aluminum hull designed for equivalent strength and stiffness to a steel hull would be about 50% thicker, but lighter by as much as 50%, and would have a 30% greater dent resistance and 13% greater resistance to rupture."[2] The corrosion resistance of aluminum is also superior to steel. Yachts up to 200 feet long are becoming popular with all 5000 series aluminum hulls. Aluminum alloy development continues in all areas.

Alloy 6061 remains the commonest manufacturing alloy in the world, and new uses seem never ending. The alloy has become dominant in yacht manufacture, and as the price of 6061 comes down, it is becoming popular in bikes again. This decade, alloy 6061 is having a revival in luxury and novelty items. Apple has used it in iPhones. However, Apple Watch has upgraded from

6000 series alloy compositions (using magnesium and silicon) to a custom 7000 series alloy that relies on zinc. The closest composition to this custom alloy is the 6061 aluminum alloy and 7075 aluminum alloy. The Apple watch has an anodized surface allowing for six different colors. Still, 6065 is cheaper and more flexible for sports application.

Alloy 6061 is also being used in TaylorMade Ghost Black Tour putters released in 2015. These golf putters use 6061 aircraft-grade aluminum face insert that saves weight over an all-stainless steel cast head. The saved weight is then redistributed to enhance the roll. A new class of compact bicycles is being made using 6061. An ice cream scoop for chefs made from forged 6061 is being offered at $50. Tactical survival pens of heat-treated 6061 are today being made by Colt and another gun company, Smith & Wesson. These pens have the hardness of steel with the weight of aluminum. Aluminum 7075 has also been made into survival pens.

Aluminum 7075 type alloys have a darker application than iPhones. Aluminum tubes have become common in the news because of their use in nuclear applications. The tubes are made from 7075 aluminum, an extremely hard alloy that made them suitable as rotors in a uranium centrifuge. Such tubes are strong enough to spin at the terrific speeds needed to convert uranium gas into enriched uranium, an essential ingredient of a nuclear bomb. For this reason, international rules prohibited Iraq from importing certain sizes of 7075 aluminum tubes. Aluminum has other key uses today. One of the most interesting uses for 7075 is in the manufacture of M16 rifles for the American military. In particular, high-quality M16 rifle lower and upper receivers, as well as extension tubes, are typically made from 7075 alloy. This alloy, of course, goes back to the secret development in Japan in the 1930s. The 7000 series is more expensive than 6061 but has better strength, making it popular in racing bikes; 6061 with its flexibility is preferred in mountain bikes. Alloy 6061 has the commercial frame market with its lower cost ($300 to $500) and smoother ride.

Aluminum's real future may be in its recycling. Two-thirds of the aluminum ever produced is still in use today. In a mere 60 days, an aluminum can is recycled, turned into a new can, and back on a store shelf. It takes about 25 cans to make a pound, and a pound of used cans gets about 55 cents. Around the United States, aluminum is recycled 50 percent of the time compared to glass and plastic, which are reprocessed less than 25 percent. Recycling has a huge potential because it requires a fraction of the energy needed to smelt from ore. Every three months, Americans throw away enough scrap aluminum to rebuild the entire U.S. commercial airplane fleet. Recycling that metal would save the energy equivalent of 16 million barrels of oil. Recycling saves 95 percent of the production of energy needed to create the metal through smelting processes. Discarding a can wastes as much energy as powering a laptop computer for 11 hours, or a television for 4 hours or the equivalent of a half a gallon of gasoline. The aluminum industry pays more than $800 million for recycled material, and every minute an average of 113,000 aluminum cans are recycled. These are powerful statistics that drive the recycling of aluminum compared to other materials.

Aluminum has become the darling of the green movement due to recycling and other factors. Replacing two pounds of steel with one pound of aluminum in a car saves 20 pounds of CO_2 emissions over the life of the car. Of course, most of the stated green advantages assume recycling. Aluminum smelting uses grid power, which is often coal-powered. Still, aluminum has continued to be favored by environmentalists over plastic and steel. Its accepted weight reduction of vehicles with aluminum offsets the percent of the energy consumption and percent of cumulative greenhouse gas emissions associated with primary aluminum production.

Just how strong the green movement is in supporting aluminum can be seen in the Danish toymaker Lego. Lego produces 60 billion plastic bricks weighing 77,000 metric tons every year. Its use of plastic had been the target of Greenpeace for a number of years. In 2015 Lego announced it will be moving to environmentally friendly aluminum blocks.

Recycling continues to bring innovation and improvements to traditional products. The aluminum bottle is one of these. The aluminum beverage bottle, made by the impact extrusion process, was introduced in the fall of 2001 by Coca-Cola as an ecological alternative to plastic bottles. The aluminum bottle is similar in shape to a traditional beverage bottle with many designs, including resealable lids and caps. A broad range of aluminum beverage bottle profiles, styles, and configurations, are available. Some, like those used by beer companies, use the twist-off cap. Consumer preferences show they want glass bottles for beer and prefer aluminum over plastic.[3] Of course, glass bottles, because of potential throwing, have been banned at sporting events.

The aluminum bottle was targeted to take on the inroads made by plastic beer bottles at sporting events. ALCOA teamed up with Pittsburgh Brewing in 2004 to produce an aluminum bottle. In 2010, Coors Light launched the 16-oz. resealable Silver Bullet Aluminum. Aluminum bottles cost more than twice as much as glass, or about a nickel more per beer. The aluminum bottle was supposed to keep beer colder for as much as 50 minutes longer than cans or glass. Many believe they have not reached that claim. These bottles contain three times more aluminum than a 12-ounce aluminum beer can. So far, the cost of the aluminum bottle has been prohibitive for wider use. The aluminum bottle has shown more success in Europe and Asia. Recycling again is the big advantage, but even here there are problems. These aluminum bottles are technically illegal in some states such as New York. It is the throwaway cap that presents a problem. Laws enacted in the 1980s intended to outlaw the old-fashioned pull-tab beer cans, which were considered a hazard to the environment and harmful to animals who ingested the discarded tabs. Aluminum bottles also have to deal with perceived health risks. The aluminum bottles use a special polymer coating to preventing leaking.

Other environmental issues continue to be addressed by the aluminum industry. ALCOA presented UltrAlloy 6020 at the Aluminium 2004 World Trade Fair and Convention in Essen. Alloy 6020 removes lead from aluminum precision-machined parts, increases machining productivity, and boosts machine shop competitiveness. For decades, lead was added to alloys to improve machinability of steel and aluminum, but it was found to be a major health hazard. There has been a significant movement since the 1970s to eliminate lead and lead-containing products in the U.S.; and abroad over the last 20 years, lead removal has been a dilemma for the majority of parts manufacturers who required lead for machining. In the 1980s, the U.S. Environmental Protection Agency (EPA) helped to phase out lead in gasoline, reduce lead in drinking water, and reduce lead in industrial air pollution. ALCOA alloys solved the dilemma for the machining and metal parts industry.

For packaging waste, one promising solution may be the use of laminated foil. For instance, packaging 65 pounds of coffee in steel cans requires 20 pounds of steel, but only three pounds of laminated packaging including aluminum foil. Such packaging also takes up less space in the landfill. The Aluminum Association's Foil Division is even developing an educational program on aluminum foil professional packaging designers in order to help inform such designers of the benefits of switching to flexible packaging. Aluminum foil also uses less energy during both manufacturing and distribution, with in-plant scrap being recycled, which makes it the favorite of the green movement.

However, things in China are much different. China is the largest single producer and consumer of aluminum in the world in 2015. Aluminum production not only consumes about 8 percent of all electric power in China, but it is also responsible for large amounts of carbon dioxide (CO_2) emissions because much of the electrical power comes from coal plants. And the electricity needed to power the industry has a multiplier effect on pollution problems because 70 percent of China's electricity comes from coal, a major contributor of CO_2 emissions. The problem is that new global pacts on pollution could hurt western aluminum production because of its large contribution to CO_2 emissions, while China remained unconstrained. With its already huge cost advantage, China would dominate the world market.

There is a potential solution for the Chinese aluminum industry. China has, at its border, a new source of alumina which will change the world industry. Vietnam, with abundant bauxite at between 5.6 and 8.3 billion tons, could become the third-largest alumina producer in the world after Guinea and Australia. Not satisfied with being just a raw materials producer, Vietnam has awarded the engineering and construction contract for two alumina projects in the central highlands to a subsidiary of state-owned Chinese Aluminum Corporation of China (Chalco). Both plants are expected to produce some 600,000 tons per year when operational, which initially will mostly go for export, but which the government hopes will form the basis for a domestic aluminum industry in the future. To do that will require a major investment in electrical energy generation. With the hydroelectric power generation sites along the Mekong River, a fierce debate is raging between environmentalists on one side and industrialists on the other as to the merits of developing what could be up to 12 hydroelectric sites along the Mekong.

The hydroelectric plants are to be spread along the 3,000-mile length as the river runs from China through four downstream countries of Thailand, Laos, and Cambodia before reaching the sea at the Mekong Delta in Vietnam. Most of the hydroelectric projects would be in China and along the Thailand/Laos border, created by the Mekong for many hundreds of miles. Vietnam would arguably be the main buyer of the power. However, at the end of the river, Vietnam will have the most to lose from the loss of silt, lower water levels, and damage to fishing stocks that would result.

Aluminum has become the metal of the environmentalists in America. There are fewer environmental issues, but a new issue of health concerns is a growing problem for the industry. These concerns are mixed, and so far, the government suggests levels found in foods are safe. The biggest source is cooking acidic foods in aluminum pots. Aluminum levels found in processed foods, and foods cooked in aluminum pots, are generally considered to be safe as well. Back around 1970, a Canadian research team found that there was a connection between aluminum in the diet and the development of Alzheimer's disease. Some studies show that people exposed to high levels of aluminum may develop Alzheimer's, but other studies have not found this to be true. The industry answer has been to produce heavy anodized aluminum cookware which reduces the aluminum going into the food. Studies continue to show mixed results even for anodized pans.

Since the 1970s, the search for a cheaper and greener manufacturing process than the Hall-Heroult has continued. Aluminum's enormous energy requirements have made many energy-poor countries such as Japan incapable of developing a primary industry. In the 1970s, research into a bacteria aluminophage that could release aluminum from clay and bauxite appeared promising. It even inspired a new generation of aluminum science fiction writers such as G.C. Edmondson and his novel *The Aluminum Man*. Edmondson's fictional aluminophage would allow scientists to challenge the aluminum monopoly, so powerful in the 1970s. In reality, the search for an aluminophage bacterium has proven far more difficult than once thought. The enthusiasm for aluminophage has failed, but another idea, alumina from coal ash, has shown promise. Moreover, several Chinese researchers are working on new technology that could produce alumina, not from traditional bauxite, but from fly ash, a waste product generated by burning coal (something China leads the world in) which can contain up to 45 percent alumina.[4]

Research is being done on exotic forms of aluminum as well. Scientists have created a transparent form of aluminum, or more correctly, aluminum oxynitride, by bombarding the aluminum with the world's most powerful X-ray laser. Transparent aluminum had been featured in the science fiction movie *Star Trek IV: The Voyage Home*, in which Scotty gave instructions for the creation of the fictional transparent aluminum material. The real material is every bit as exotic as the one created by Scotty. It is 4 times harder than fused silica glass, 85 percent as hard as sapphire. As a transparent armor material, it provides a bulletproof product with far less weight and thickness than traditional bulletproof glass.

Another amazing aluminum material developed by the researchers is lightweight aluminum foam that expands. The material is sandwiched between two sheets of steel and sealed with heat, resulting in a very lightweight and tough building material. The finished foam product has a pore cell structure that is not only ultra-light in weight, but provides improved energy absorption, is fire and heat retardant as well as sound resistant. Under stress, the material bends but doesn't break, making it ideal for ships traveling even through dangerous ice-packed waters.

Aluminum is making a comeback to its 1930s place in art. That aluminum art movement is focused in Australia, the world's keystone to aluminum production. Australian Marc Newson's aluminum "Lockheed Lounge" chair was exhibited in 1986 at the Roslyn Oxley Gallery in Sydney. The "Lockheed Lounge" has a striking resemblance to Art Deco of the 1930s. It received international exposure when Madonna used it in her 1988 "Rain" video, and it was purchased for the Paramount Hotel in New York in 1990. The sensuous curves and biomorphic forms were also used in Newson's designs for the aluminum "Orgone chair" of 2000. The Lockheed Lounge has become a signature piece for Newson's design firm that now has offices around the world. Newson has created a renaissance in industrial design. He was involved with the design of the Apple Watch. Newson has become known as a designer who works with aluminum in any scale. His company has provided designs for watches, bicycles, and household objects, and has grown to include special collectors' items as well as prestige designs for corporate clients. One of his limited series of 15 Lockheed Lounge Chairs fetched $3.7 million in 2015. In 2014, he produced an anodized aluminum as he became vice-president of design for Apple. Newson has inspired a new generation and set a style for a number of Australian artists.

Aluminum cookware of all types is very popular again. Nonstick, scratch-resistant anodized aluminum cookware is a good choice, as is heavy cast anodized aluminum. Anodized aluminum surfaces are nonstick and eliminate the health hazards of Teflon-type coatings. Remember, anodized aluminum produces a thick skin of aluminum oxide used in the 1920s to protect aircraft skin. It is literally an inert coating of ceramic aluminum oxide that cannot be reduced to aluminum by cooking foods. The hard surface is easy to clean. It is sealed so aluminum cannot get into food. The use of a nonstick surface further protects the aluminum and the consumer. The Alzheimer's Association reports that using anodized aluminum cookware is not a major risk for the disease. Uncoated and non-anodized aluminum cookware is a greater risk. This type of cookware can easily melt on the stove as well. It can cause burns if it gets too hot. Still, research has shown that the amount of aluminum this cookware leaks into food is very small.

The future is as bright for aluminum as it was in 1899.

Chapter Notes

Preface
1. Bernard Jaffe, *Chemistry Creates a New World* (New York: Thomas Crowell), 1957.

Chapter 1
1. Mini Sheller, *Aluminum Dreams* (Cambridge: MIT Press, 2014) p. 47.
2. Ibid.
3. S. I. Venetsky, *Tales about Metals* (Moscow: Mir Publishers, 1978), p. 29.
4. John Emsley, *Nature's Building Blocks* (Oxford: Oxford University Press, 2001), p. 22.
5. John Emsley, *Nature's Building Blocks* (Oxford: Oxford University Press, 2001), p. 22.
6. Harry Holmes, "Fifty Years of Aluminum," *The Scientific Monthly*, Vol. 42, No. 3, March 1936, pp. 236–239.
7. *Household Words*, December 13, 1856, as referenced by European Aluminum Association.
8. S. I. Venetsky, *Tales about Metals* (Moscow: Mir Publishers, 1978), p. 25.
9. Walter James Miller, *The Annotated Jules Verne: From the Earth to the Moon* (New York: Thomas Y. Crowell, 1978), p. 46.

Chapter 2
1. P. T. Stroup, "Hall, Heroult, Aluminum and Fused Salt Electrochemistry," *Transactions of the ASM*, Vol. 62, 1969, pp. 1045–1078.
2. J. A. Price, "Aluminum," supplement in *Scientific American*, January 18, 1886.
3. Alfred Cowles, "Production of Aluminum and its Alloys in Electric Furnace," *Proceedings of the Franklin Institute*, January 1886.
4. Alfred Cowles, *The True Story of Aluminum* (Chicago: Henry Regnery, 1958), p. 91.
5. Alfred Cowles, *The True Story of Aluminum* (Chicago: Henry Regnery, 1958), p. 92.
6. Alfred Cowles, *The True Story of Aluminum* (Chicago: Henry Regnery, 1958), p. 36.
7. Donald Wallace, *Market Control in the Aluminum Industry* (Cambridge: Harvard University Press), p. 508.
8. Harry Holmes, "The Story of Aluminum," *Journal of Chemical Education*, February 1930, p. 2.
9. Harry Holmes, "Fifty Years of Aluminum," *The Scientific Monthly*, Vol. 42, No. 3, March 1936, pp. 236–239.
10. British Patent No. 1214, May 13, 1861, filed by Thomas Bell—an agent of Louis Le Chatelier.
11. Warren Peterson and Ronald Miller, editors, *Hall-Heroult Centennial: First Century of Aluminum Process Technology 1886-1986* (Warrendale: Metallurgical Society, 1986), pp. 115–119.
12. French Patent 175, 711 by Paul Heroult, April 23, 1886; American Patent 400,7666 by Charles hall April 2, 1889.

Chapter 3
1. *Frank Fanning Jewett: The Beloved Teacher*, Booklet of Oberlin College, 1926, p. 33.

2. George David Smith, *From Monopoly to Competition* (Cambridge: Cambridge University Press, 1988), p. 88.
3. Holmes, "A Great Pupil and a Great Discovery-Both Supported by a Great Teacher," *Science*, Vol. 83, No. 2147, Feb. 21, 1936, pp. 175–177.
4. *Frank Fanning Jewett: The Beloved Teacher*, Booklet of Oberlin College, 1926, p. 34.
5. *Frank Fanning Jewett: The Beloved Teacher*, Booklet of Oberlin College, 1926, p. 34.
6. Holmes, "A Great Pupil and a Great Discovery-Both Supported by a Great Teacher," *Science*, Vol. 83, No. 2147, Feb. 21, 1936, pp. 175–177.
7. "Perkin Medal Award," *Journal of Industrial Engineering and Chemistry*, March 1911, pp. 145–51.
8. Junius Edwards, *The Immortal Woodshed*, (New York: Dodd, Mead, 1955), p. 140.
9. Rosamond McPherson Young, *Made of Aluminum: A Life of Charles Hall* (New York: David McKay, 1965), p. 66.
10. Claudia Flavell, "Turning a Rarity into a Commodity," from Chemical Engineers that Changed the World, *TCE Today*, tcetoday.com, June 2013.
11. Adolphe Minet, *The Production of Aluminum and its Industrial Uses* (New York: Wiley, 1905), pp. 91–96.

Chapter 4

1. Alfred Cowles, *The True Story of Aluminum* (Chicago: Henry Regnery, 1958), p. 86.
2. Alfred Cowles, *The True Story of Aluminum* (Chicago: Henry Regnery, 1958), p. 78.
3. "Pittsburgh and the Pittsburgh Spirit," Pittsburgh Chamber of Commerce, 1927, p. 184.

Chapter 5

1. George David Smith, *From Monopoly to Competition* (Cambridge: Cambridge University Press, 1988), p. 80.
2. David Cannadine, *Mellon: An American Life* (New York: Alfred A. Knopf, 2006), pp. 102–122.

3. Harvey O'Conner, *Mellons Millions* (New York: John Day, 1933), p. 83.
4. James Rock, "A Growth Industry: The Wisconsin Cookware Industry," *The Wisconsin Magazine of History*, Vol. 55, No. 2, 1971, pp. 86–90.
5. Alfred Cowles, *The True Story of Aluminum* (Chicago: Henry Regnery, 1958), pp. 98–106.
6. Alfred Cowles, *The True Story of Aluminum* (Chicago: Henry Regnery, 1958), p. 105.
7. *Engineering News*, Volume L, 1903, p. 390.

Chapter 6

1. *Manufacturer and Builder*, November 1888, Volume 20, Issue 11.
2. Quentin Skrabec, *H. J. Heinz: A Biography* (Jefferson: McFarland, 2009), pp. 228–277.
3. Charles C. Carr, *ALCOA: An American Enterprise* (New York: Rinehart, 1952), p. 62.
4. George David Smith, *From Monopoly to Competition* (Cambridge: Cambridge University Press, 1988), p. 95; a quote from a letter of Hunt to Hall 1897 presently in ALCOA archives at Western Pennsylvania History Museum.
5. George Bachus, "Background and Early History of a Company Town: Bauxite, Arkansas," *The Arkansas Historical Quarterly*, Vol. 27, No. 4, Winter, 1968, pp. 330–357.

Chapter 7

1. "Golden Jubilee of Aluminum," *Science*, Vol. 83, No. 2151, March 20, 1936.
2. Burton Hersh, *The Mellon Family* (New York: William Morrow, 1978) p. 148.
3. Martin Perry, "Forward Integration of ALCOA: 1888–1930," *The Journal of Industrial Economics*, Vol. 29, No. 1, 1980, p. 37.
4. James Rock and Brian Peckham, "Depression, and War: The Wisconsin Aluminum Cookware Industry," *Wisconsin Magazine of History*, Vol. 73, No. 3, Spring, 1990, p. 213.
5. Tudor Van Hampton, "For Cars, Aluminum Is a Back to the Future Metal," *New York Times*, Feb. 14, 2014.

6. Martin Perry, "Forward Integration of ALCOA: 1888–1930," *The Journal of Industrial Economics*, Vol. 29, No. 1, 1980, p. 42.

7. Dietrich Altenpohl, *Aluminum: Technology, Applications, and Environment* (Washington: Aluminum Association, 1998), p. 212.

8. Margaret Graham, "R&D and Competition in England and the United States: The Case of the Aluminum Dirigible," *Business History Review*, Vol. 16, 1949, p. 267.

9. Today Oberlin College is the home of Hall's original pieces of aluminum as well as his archives. An aluminum statue of Hall is also located there.

10. David Cannadine, *Mellon: An American Life* (New York: Alfred A. Knopf, 2006), pp. 223–30.

Chapter 8

1. George David Smith, *From Monopoly to Competition* (Cambridge: Cambridge University Press, 1988), p. 77.

2. Clive Edwards, "Aluminium Furniture, 1886–1986: The Changing Applications and Reception of a Modern Material," *Journal of Design History*, Vol. 14, No. 3 (2001), pp. 207–225.

3. Martin Perry, "Forward Integration of ALCOA: 1888–1930," *The Journal of Industrial Economics*, Vol. 29, No. 1, 1980, p. 37.

Chapter 9

1. Burton Hersh, *The Mellon Family* (New York: William Morrow, 1978), p. 385.

2. Ann Spackman, "The Role of Private Companies in the Politics of the Empire: A case Study of Bauxite and Diamond Companies in Guyana in early 1920s." *Social and Economic Studies*, Vol. 24, No. 3, 1975, p. 350.

3. Duncan Campbell, *Global Mission: The Story of Alcan*, (Ontario Publishing Company Limited, 1985), pp. 9–12.

4. National Newcomen Dinner of The Newcomen Society of England in New York. New York, United States of America, on April 26, 1951.

5. Martin Perry, "Forward Integration of ALCOA: 1888–1930," *The Journal of Industrial Economics*, Vol. 29, No. 1, 1980, p. 112.

6. "Muscle Shoals," *Boston Ideas*, January 1 to October 1, 1927, Benson Ford Research Center, Dearborn, Michigan, Linear files, Muscle Shoals.

7. Littell McClung, "What Can Henry Ford Do With Muscle Shoals," *Illustrated World*, April 1922, Vol. 27, no. 2, p. 186.

Chapter 10

1. Carl Meyerhuber, "Organizing ALCOA: The Aluminum Workers' Union in the Pennsylvania's Allegheny Valley, 1900–1971," *Pennsylvania History*, Vol. 48, No. 3, July 1981, pp. 195–219.

2. George Bachus, "Background and Early History of a Company Town: Bauxite, Arkansas," *The Arkansas Historical Quarterly*, Vol. 27, No. 4, Winter, 1968, pp. 330–357.

3. Russell Parker, "The Black Community in a Company Town: ALCOA Tennessee, 1919–1939." *The Tennessee Historical Quarterly*, Vol. 37, No. 2, 1976, pp. 203–221.

4. David Duggan and George Williams, *ALCOA* (Charleston: Arcadia Publishing, 2011), p. 9.

5. David Duggan and George Williams, *ALCOA* (Charleston: Arcadia Publishing, 2011), p. 1.

6. Harper Barnes, *Never Been a Time: The 1917 Race Riot That Sparked the Civil Rights Movement*, (New York: Walker, 2008) pp. 20–50.

7. Duncan Campbell, *Global Mission: The Story of Alcan* (Ontario Publishing Company Limited, 1985), p. 128.

8. Charles C. Carr, *ALCOA: An American Enterprise* (New York: Rinehart, 1952), p. 195.

Chapter 11

1. Carl I. Meyerhuber Jr., "Organizing ALCOA," *Pennsylvania History: A Journal of Mid-Atlantic Studies*, Vol. 48, No. 3 (July 1981), pp. 195–197.

2. Carl I. Meyerhuber Jr., "Organizing ALCOA," *Pennsylvania History: A Journal of Mid-Atlantic Studies*, Vol. 48, No. 3 (July 1981), pp. 201–202.

3. Carl I. Meyerhuber Jr., "Organizing ALCOA," *Pennsylvania History: A Journal of Mid-Atlantic Studies*, Vol. 48, No. 3 (July 1981), p. 199–200.

4. Russell Parker, "The Black Community in a Company Town: ALCOA Tennessee, 1919–1939." *The Tennessee Historical Quarterly*, Vol. 37, No. 2, 1976, p. 218.

Chapter 12

1. David Cannadine, *Mellon: An American Life* (New York: Alfred A. Knopf, 2006), p. 516.
2. Charlotte Muller, "The Aluminum Monopoly and the War," *Political Science Quarterly*, Vol. 60, No. 1, March 1949, p. 15.
3. R. Buckminster Fuller, *Critical Path* (New York: St. Martins' Griffin, 1981), p. 98.
4. George David Smith, *From Monopoly to Competition* (Cambridge: Cambridge University Press, 1988), p. 199.
5. Duncan Campbell, *Global Mission: The Story of Alcan* (Ontario Publishing Company Limited, 1985), p. 231.
6. "A Reporter at Large-Thurman Arnold's Biggest Case," *The New Yorker*, January 1942.
7. United States vs. Aluminum Company of America, 44 F Supplement 97, pp. 308–309.
8. United States vs. Aluminum Company of America, 44 F Supplement 97, p. 306.
9. Ayn Rand, *Capitalism: An Unknown Ideal* (New York: New American Library, 1967), p. 107.

Chapter 13

1. "Aluminum Reborn," *Fortune*, May 1946.
2. Charles R. Geisst, *Monopolies in America: Empire Builders & Their Enemies form Jay Gould to Bill Gates* (Oxford: Oxford University Press, 2000), p. 187.
3. Theodore Gray, "The Amazing Rusting Aluminum," *Popular Science*, September 22, 2004.
4. Editorial, *The New York Times*, October 17, 1945.

Chapter 14

1. Tom Martens, "Disneyland, and to Tomorrow land's forgotten Pig," *Los Angeles Times*, July 15, 2015.

2. Dennis Doordan, "Aluminum Designers and the American Aluminum Industry," *Design Issues*, Vol. 9, No. 2 (Autumn 1993), p. 47.
3. Dennis Doordan, "Aluminum Designers and the American Aluminum Industry," *Design Issues*, Vol. 9, No. 2 (Autumn 1993), p. 38.

Chapter 15

1. Speech by David Macintyre, February 12, 1959. Heinz History Center Archives, MSS 282, Box 11, Folder 7.
2. Mini Sheller, *Aluminum Dreams* (Cambridge: MIT Press, 2014) p. 156.
3. Robert E. Sanders, Jr., "Technology Innovation in Aluminum Products," *JOM*, Vol. 53, 2001, pp. 21–25.
4. Jeffery Schnapp, "The Romance of Caffeine and Aluminum," *Critical Inquiry*, Vol. 28, No. 1, (Autumn 2001), p. 258.
5. Suzette Worden, "Aluminium and Contemporary Australian Design: Materials History, Cultural and National Identity," Journal of Design History Vol. 22 No. 2, 2009.
6. Lloyd Steven Seiden, *Buckminster Fuller's Universe* (New York: Perseus Publishing, 1989), pp. 2–123.
7. Athena V. Lord, *Pilot for Spaceship Earth* (New York: Macmillan, 1978), pp. 50–61.
8. Alden Hatch, *Buckminster Fuller at Home in the Universe* (New York: Crown Publishers, 1974) pp. 288–302.

Chapter 16

1. Sarah Nichols, *Aluminum by Design* (Pittsburgh: Carnegie Museum of Art, 2000), p. 104.
2. John Lauber, "And It Never Needs Painting: The Development of Residential Aluminum Siding," *Association for Preservation Technology International*, Vol. 31, No. 2, 2000, pp. 17–24.
3. Subodh K. Das and J. Gilbert Kaufman, "Aluminum Alloys for Bridges," The Minerals, Metal, and Materials Society, 2007.
4. Sarah Nichols, *Aluminum by Design* (Pittsburgh: Carnegie Museum of Art, 2000), p. 143.

5. Editorial, The Economic Weekly, London, January 17, 1950.

Chapter 17

1. ALCOA Archives, Heinz History Center, MSS 282, Box 17, Folder 5.
2. Drew Middleton, "Falklands Aftermath: A Naval Debate," New York, July 10, 1982.

Chapter 18

1. Henry Petroski, *The Evolution of Useful Things* (New York: Vintage Books, 1992), p. 201.

Chapter 19

1. Marylynn Uricchio, "An art-filled aluminum abode," *Pittsburgh Quarterly,* Fall 2009.
2. Roger Ganem, "The Aluminum Shaft Dilemma," *Golfdom,* February 1968, p. 37.
3. John Pearley Huffman, "How the Chevy Vega Nearly Destroyed GM," *Popular Mechanics,* October 19, 2010.

Chapter 20

1. Mini Sheller, *Aluminum Dreams* (Cambridge: MIT Press, 2014), p. 140.
2. Betsy Brown, "Officials, in Visits to Factories, Explore Modular Homes." *The New York Times,* 4/20/86.
3. Barry Rehfeld "Even some Contractors are Choosing Modular Homes." *The New York Times* 9/30/06.

Chapter 21

1. "Out of Siberia, a Russian Way to Wealth," *New York Times,* August 20, 2006.
2. Carol J. Williams, "Jamaica: Dust-Up Swirls Around Key Jamaica Industry," *Los Angeles Times,* October 25, 2004.
3. Simon Denyer, "In China's Inner Mongolia, mining spells misery for traditional herders," *Washington Post,* April 5, 2015.

Chapter 22

1. "Chevy ad escalates pickup war, puts steel vs. aluminum issue on the table," *Automotive News,* July 16, 2015.
2. Michael Skillingberg, "Aluminum at Sea: Speed, endurance and affordability," *Marine Log,* May 2005, pp. 28–32.
3. Jim George, "Survey: Perception huge in beverage packaging," *Packaging World,* October 7, 2006.
4. Andy Home, "Bauxite supply no hindrance to China's aluminum boom," *Reuters London,* June 9, 2015.

Bibliography

Archives
ALCOA Papers—Historical Society of Western Pennsylvania
Andrew Mellow Papers—National Gallery of Art, Washington, D. C.
Carnegie Library Oakland—Pennsylvania Room
Frick Papers and Frick Collection, University of Pittsburgh
Mellon Bank Papers—Historical Society of Western Pennsylvania
Thomas Mellon Papers—National Gallery of Art, Washington, D. C.
William Martin Papers—Archives of Industrial Society, University of Pittsburgh Library
William McKinley Papers—William McKinley Presidential Library, Canton, Ohio

Books and Articles
Altenpohl, Dietrich. *Aluminum: Technology, Applications, and Environment.* Washington: Aluminum Association, 1998.
Bachus, George. "Background and Early History of a Company Town: Bauxite, Arkansas," *The Arkansas Historical Quarterly*, Vol. 27, No. 4, Winter, 1968.
Barnes, Harper. *Never Been a Time: The 1917 Race Riot That Sparked the Civil Rights Movement*, New York: Walker, 2008.
Campbell, Duncan. *Global Mission: The Story of Alcan.* Ontario Publishing Company Limited, 1985.
Cannadine, David. *Mellon: An American Life.* New York: Alfred A. Knopf, 2006.
Carr, Charles. *ALCOA: An American Enterprise.* New York: Rinehart, 1952.
Cowles, Alfred. *The True Story of Aluminum.* Chicago: Henry Regnery, 1958.
Cowles, Eugene. "Production of Aluminum and Its Alloys in Electric Furnace," *Proceedings of the Franklin Institute*, January 1886.
Doordan, Dennis. "Aluminum Designers and the American Aluminum Industry." *Design Issues*, Vol. 9, No. 2 (Autumn, 1993), pp. 44–50.
Duggan, David, and George Williams. *ALCOA.* Charleston: Arcadia Publishing, 2011.
Edwards Junius, *The Immortal Woodshed.* New York: Dodd, Mead, 1955.
Emsley, John. *Nature's Building Blocks.* Oxford: Oxford University Press, 2001.
Flavell, Claudia. "Turning a Rarity into a Commodity," from Chemical Engineers that Changed the World. *TCE Today,* tcetoday.com, June 2013.
Frank Fanning Jewett: The Beloved Teacher, Booklet of Oberlin College, 1926.
Fuller, Buckminster. *Critical Path.* New York: St. Martins' Griffin, 1981.
Ganem, Roger. "The Aluminum Shaft Dilemma," *Golfdom.* February 1968.

Geisst, Charles. *Monopolies in America: Empire Builders and Their Enemies from Jay Gould to Bill Gates.* Oxford: Oxford University Press, 2000.
Graham, Margaret B.W. "R&D and Competition in England and the United States: The Case of the Aluminum Dirigible," *Business History Review*, Vol. 16, 1949.
Hatch, Alden. *Buckminster Fuller at Home in the Universe.* New York: Crown Publishers, 1974.
Hersh, Burton. *The Mellon Family.* New York: William Morrow, 1978.
Holmes, Harry. "Fifty Years of Aluminum," *The Scientific Monthly*, Vol. 42, No. 3, March 1936.
Holmes, Harry. "A Great Pupil and a Great Discovery—Both Supported by a Great Teacher." *Science*, Vol. 83, No. 2147, Feb. 21, 1936.
Holmes, Harry. "The Story of Aluminum." *Journal of Chemical Education.* February 1930.
Home, Andy. "Bauxite supply no hindrance to China's aluminum boom." *Reuters London*, June 9, 2015.
Huffman, John. "How the Chevy Vega Nearly Destroyed GM." *Popular Mechanics*, October 19, 2010.
Lauber, John. "And It Never Needs Painting: The Development of Residential Aluminum Siding," *Association for Preservation Technology International*, Vol. 31, No. 2, 2000.
Lord, Athena. *Pilot for Spaceship Earth.* New York: Macmillan, 1978.
McClung, Littell. "What Can Henry Ford Do With Muscle Shoals." *Illustrated World*, April 1922, Vol. 27, no. 2.
Meyerhuber, Carl. "Organizing ALCOA: The Aluminum Workers' Union in the Pennsylvania's Allegheny Valley, 1900–1971." *Pennsylvania History*, Vol. 48, No. 3, July 1981, pp. 195–219.
Miller, Walter. *The Annotated Jules Verne: From the Earth to the Moon.* New York: Thomas Y. Crowell, 1978.
Minet, Adolphe. *The Production of Aluminum and Its Industrial Uses.* New York: Wiley, 1905.
Muller, Charlotte. "The Aluminum Monopoly and the War," *Political Science Quarterly*, Vol. 60, No. 1, March 1949, pp. 14–18.
Nichols, Sarah. *Aluminum by Design.* Pittsburgh: Carnegie Museum of Art, 2000.
Parker, Russell. "The Black Community in a Company Town: ALCOA Tennessee, 1919–1939." *The Tennessee Historical Quarterly*, Vol. 37, No. 2, 1976, pp. 203–221.
Perry, Martin. "Forward Integration of ALCOA: 1888–1930." *The Journal of Industrial Economics*, Vol. 29, No. 1, 1980. pp. 37–53.
Peterson, Warren, and Ronald Miller, editors, *Hall-Heroult Centennial: First Century of Aluminum Process Technology 1886–1986.* Warrendale: Metallurgical Society, 1986.
Petroski, Henry. *The Evolution of Useful Things*, New York: Vintage Books, 1992.
Price, J. A. "Aluminum," supplement in *Scientific American*, January 18, 1886.
Rand, Ayn. *Capitalism: An Unknown Ideal.* New York: New American Library, 1967.
Rock, James. "A Growth Industry: The Wisconsin Cookware Industry." *The Wisconsin Magazine of History*, Vol. 55, No. 2, 1971.
Rock, James, and Brian Peckham, "Recession, Depression and War: The Wisconsin Aluminum Cookware Industry," *Wisconsin Magazine of History*, Vol. 73, No. 3, Spring, 1990, pp. 202–223.
Sanders, Robert, Jr. "Technology Innovation in Aluminum Products." *JOM*, Vol. 53, 2001, pp. 21–25.
Schnapp, Jeffery. "The Romance of Caffeine and Aluminum." *Critical Inquiry*, Vol. 28, No. 1 (Autumn, 2001), pp. 244–269.
Seiden, Lloyd. *Buckminster Fuller's Universe.* New York: Perseus Publishing, 1989.
Sheller, Mini. *Aluminum Dreams.* Cambridge: MIT Press, 2014.
Skillingberg, Michael. "Aluminum at Sea: Speed, Endurance and Affordability." *Marine Log*, May 2005, pp. 28–32.
Skrabec, Quentin. *H. J. Heinz: A Biography.* Jefferson: McFarland, 2009.

Smith, George. *From Monopoly to Competition*. Cambridge: Cambridge University Press, 1988.
Spackman, Ann. "The Role of Private Companies in the Politics of the Empire: A Case Study of Bauxite and Diamond Companies in Guyana in Early 1920s." *Social and Economic Studies*, Vol. 24, No. 3, 1975, pp. 240–370.
Stroup, P.T. "Hall, Heroult, Aluminum and Fused Salt Electrochemistry." *Transactions of the ASM*, Vol. 62, 1969.
Uricchio, Marylynn. "An Art-filled Aluminum Abode." *Pittsburgh Quarterly*, Fall 2009.
Van Hampton, Tudor. "For Cars, Aluminum Is a Back to the Future Metal." *New York Times*, Feb. 14, 2014.
Venetsky, S. I. *Tales about Metals*. (Moscow: Mir Publishers, 1978).
Wallace, Donald. *Market Control in the Aluminum Industry*. Cambridge: Harvard University Press, 1938.
Young, Rosamond McPherson. *Made of Aluminum: A Life of Charles Hall*. New York: David McKay, 1965.

Index

Abramovich, Roman 212–214
Adolph Coors Company 192–194
Africa 65
Aging of aluminum 71
Aircraft 139–144
Aircraft parts 88–89, 140–141
AIROH house 151–152
Airships 161–165
Airstream 176–178
ALCAN (Aluminium Company of Canada) 49, 58, 95–103, 127–131, 136–138, 147–158, 185, 209, 214
Alclad 89, 143–144
ALCOA (Aluminum Company of America) 1, 11, 22, 38–42, 46–49, 63–69; Alcan split 76–78; anti-trust cases 124–133; black labor 108–112; cartel behavior 64–67, 93–94, 127–130; city building 103–110; cruises 160–163; labor relations 103–112, 114–123; market strategy 186–187; Mexican labor 107–110; monopoly of 63–67, 94, 100, 125–127, 130–133, 144–146; organization 154–156; racial relations 106–108; research department 154; wages 122–123
Alcoa, Tennessee 82–83, 107–110, 120–122
ALCOA Building 1, 170, 174–175
ALCOA Combination Process 138
ALCOA Research Laboratory (New Kensington) 89–91
Alliance Aluminium Compagnie 127–128
Alumina (Aluminum Oxide) 6, 22–25, 60–61
Aluminium (European spelling) 10
Aluminum A.G. 65
Aluminum Alloys: alloy A-390, 2; alloy 24S 141; alloy 17S 72, 83, 140; alloy 25S (2025) 88; alloy 398; alloy 61S 141; alloy 75S 142; armetale 199; heat treating of 89; lithinum alloys 70, 220–221; series 1000, 139; series 2000, 139, 180; series 2017, 70; series 2020, 220; series 3000, 140, 194; series 4000, 140; series 5000, 72; series 5086; series 6000, 70, 208–209; series 6061, 141, 160, 220–221; series 6065, 141; series 6069, 208; series 7000, 142, 222; series 7050, 207; series 7075, 199, 222; series 8000, 206; series 8090, 220–221
Aluminum Bronze 19–22, 33–34, 36, 45, 53
Aluminum Casting Company 73–74
Aluminum Cooking Utensil Company 48–49
Aluminum Industries (AIAG) 49
Aluminum Intensive Vehicle Program 209
"Aluminum killed steel" 89
Aluminum Ladder Company 181–182
Aluminum Seal Company 89
Aluminum Vehicle Technology 209
Aluminum Workers of America (CIO) 122–123
Alzheimer's Disease 225–227
American Chemical Society 10
American Development Company 93
American Federation of Labor (AFL) 102–103, 115–120
American Society of Materials 14
Anaconda Copper 148
Anodizing 158–159, 182–183, 225
Apollo spacecraft 13, 195–196
Apple Iphone 176
Appliances 180
Applications (general) 19–20, 66–67, 72–74, 143
Arkansas 60–62, 138
Armetale 199
Art Deco 3, 84, 90, 147, 156–164
Arvida, Canada 94, 109–112, 146
Arvida Bridge 180–181
ASM International Dome 176
Assembly lines 143, 172, 176, 204
Audi A 8, 209

239

August, Wendell 165–166
Australian aluminum industry 167, 214–216
Avery Stamping Company 53
Awning 147–148, 203

Badin, Adrien 74
Badin, North Carolina 74–75
Baltimore plant 74
Baseball bats 206–207
Bauxite 5–6, 22–25, 58, 60–61, 65, 73–74, 94, 97, 109–110, 127, 135, 137, 214–216
Bauxite, Arizona 64
Bauxite, Arkansas 61–64, 93, 106–107
Bayer, Karl 61–63
Bayer Process 61–64
Berea College 76
Berzelius, Jons 22–23
Bessemer, Henry 17
Beverage bottles 223–224
Bialetti, Alfonso 166–167
Bicycles 67–68, 72, 86, 207–209
Big Six 187
Black Diamond Steel 38–45
Boats 221
Boeing 707, 184
Boeing B-9 140
Boeing B-29 143
Boeing 247 140
Borrani, Carlo 86
Bowlus, William 176–177
Bows (archery) 183
Bradley, Charles 54–55
Breuhaus, Fritz 162
Bridges 180–181
British Aluminium 58–59, 62–64, 74, 97–99, 151–154, 185–186
British de Havilland Comet (airplane) 184
Brush, Charles 18, 33–36
Brush Electric Company 36–37
Buckeye Iron and Brass 68
Buffalo plant 86–88
Bugatti, Ettore 202
Buildings 159–165, 174–176
Bunsen, Robert von 10–11
Burrell Improvement Company 51
Busses 188–189
Byam, Wally 176–177, 182

Cable 66–67
Caboni, V. 182
Cabris, Sam 181–182
Caffey, Francis (judge) 128–130
Canadian aluminum industry 93–103, 109–112
Canning industry 192–196
Cans 192–196
Canteen 45, 81
Car bodies 100–101, 167, 218
Carbon Steel Company 42–43

Carborundum Company 56–58
Carbothermic Process 18
Carnegie, Andrew 59, 105–106
Carnegie Steel 43
Cars 48, 68–70, 102, 167–169, 217–219
Casting and cast parts 14–15, 73–74, 83, 88–89, 165, 180, 182–183, 200–204
Castner, Alexander 15
Chalco Company (Aluminum Company of China) 213–215
Chatelier, Louis 23–24
Chernoi, Lev 212–213
Chernyshevsky, Nikolai 12–13
Chevy Vega 2, 200–203
Chicago World's Fair of 1893 52
China 9, 224
Chinese aluminum industry 95, 167
Christmas trees 204
Christofle (famous jeweler) 12
Clapp, George 37–44
Clay 5
Club Aluminum Utensil Company 83
Coinage 79, 203–204
Cole, Romaine 38–40
Combs 53
Comet (train) 169
Commercial airliners 184–185
Congress of Industrial Organization 116–120
Consent Decree of 1912 65–66
Cooking utensils and cookware 46–49, 53, 66–67, 73–74, 83, 91–92
Coors *see* Adolf Coors Company
Copper 6, 8, 19–20, 67–68
Corundum 6
Cowles, Alfred 18–19, 36–37
Cowles, Eugene 18–20
Cowles brothers 18–19, 21, 30, 36
Cowles Brothers Company 18–25, 38–37, 53–55
Cowles Electric Smelting Company 18–20
Cryolite 22–25, 55, 61–62, 136–138

Davis, Arthur 43–49, 59, 76–78, 93, 97, 104, 108–109, 127–130, 152–154
Davis, Edward 76, 96–99, 127–130, 144–146, 154–155
Davis, Nathanael 156–157
Davy, Humphry 9–10
DC-2 140
DC-3 184
DC-8 184
DC-10 137–138
Defense Plant Corporation 135–138, 144
de Havilland Comet 7
Deripaska, Oleg 212–214
Design 156–165
Deville, Henri Sainte Claire 11–13, 23, 32–33, 49

Dickens, Charles 10, 12–14
Diners 173
Disneyland 150–151
Dow Chemical 99–101
Drawn and iron canmaking 193–194
Duke, James 93–94
Duke Power Company 93–94
Duralumin 69–77, 80–81, 85–86, 130, 134–136, 139–140, 169, 177, 182, 219
Dymaxion car 69, 150, 167, 169–170
Dymaxion house 167, 172

Edison, Thomas 27–32, 53–55
Electric Dynamo 18–24, 33–35
Electric fires 205–206
Electric furnace 21–23
Electroplating 29–30
Elektro (robot) 165–166
Engines and engine parts 67–68, 200–203
Environmental issues 216–218, 224–226
Equitable Building (Portland) 175
Explosives 74, 83

Falls City Beer 194–195
Fatigue 190
Federal trade Commission 77, 82, 97
Ferrari 209
First Center for Visual Arts 3, 164
Fluorspar 23–24
Foil 45–46, 48, 91, 127, 150–151
Folding chairs 181
Ford, Henry 100–103, 132
Ford F-150 truck 141, 217
Ford Motor Company 140–144, 207
Ford Trimotor airplane 140, 167
Fordney McCumber Tariff 80–81
Forged 201–203
Frary, Francis 83–84
Frey, Albert 171–172
Frick, Henry Clay 45–47, 52, 80, 104
Frishmuth, William 14–15
Fuel 195–196
Fuller, Buckminster 2, 7, 126, 159–160, 170–174
Furniture 85–88, 158

General Agreement on Tariffs and Trade 187–188
Geodesic domes 2, 175–176
Georgia 60–61
German aluminum industry 68–74, 83, 100, 136–138
Germany 136–137
Gibbs, Oliver 28–29
Giftware 198–200
Gold 7
Golf equipment 199–201
Goodyear Zeppelin Corporation 169

Gray, Elisha 27–28, 31–32, 35
Great Depression 114–118, 129–130, 172
Greenland 136–137
Guild, Lurelle 165–166
Gulf oil 80
Gunnison, Foster 172–173
Gutters 197–198
Guyana (British Guiana) 94, 97, 137–138

Hall, Charles 3, 17–31, 52, 58–60, 72–76, 98
Hall, Julia 30–31
Hall-Heroult Process 17–25, 203, 217
Hall Process 1, 85
Hanna, Mark 80
Hannibal, Ohio plant 149
Harding, Warren 80–81
Harvard University 28–29
Harvey Aluminum Company 148–149
Haskell, George 94, 124–125
Haskell Manufacturing 124–126, 130
Hawaiian Village Hotel 175–176
Haymarket Riot 118
Head, Howard 183
Heroult, Paul 8, 17, 20–25, 30–34, 54
Hindenburg (airship) 161–162
Hoover, Herbert 80–81, 124
Hoover Building 164
Hornbostel, Henry 90–91
Horseshoes 44
Housing 151–152, 169–174, 178–180, 197–198
Hunsiker, Millard 42–43
Hunt, Alfred 38–46, 58–60, 73–76
Hunt, Roy 76, 96–103, 145–147, 152–153
Hunt mansion (Elmhurst) 198

IG Farbenindustrie 100
Imperial Kenmore vacuum cleaner 89
Indonesia 215–216
Interlocken School of Arts 88
Iron City Beer 193–194
Italian aluminum industry 168–170, 182

Jaguar 209, 217–218
Jamaica 5, 138–139, 214–215
Japanese aluminum industry 79, 139–143, 210–211
Japanese Zero (aircraft) 142, 176
Javel Chemical Works 12
Jewett, Frank 28–34
Jones, Jesse 135–136
Junkers, Hugo 73
Junkers fighters 72–73, 86, 140, 177

Kaiser, Henry 149–150, 175
Kaiser Aluminum 134, 138–139, 145, 149–151, 175–179, 194–195
Kelly, Thomas 17, 38
Kensington Ware 164–165

Kingston plant (Ontario) 151–152
Knights of Labor 118
Knox, Philander 64
Koenig Manufacturing Company 53
Krueger Brewing Company 192

Ladders 181–182
Landau (Pierce-Arrow RV) 179–180
Lanham Act 172–173
Lash, Howard 42
League of Nations 80
Les Baux (France) 5
Liberty engine 73
Local 18356, 119–122
Lowery, Grosvenor 53–54
Lu-Mi-Num Bicycle 67

M-10000 (Union Pacific train) 169
Magnesium 99–101
Makhmudov, Iskander 212–213
Marconi, Guglielmo 17
Martin B-10, 140
Mason jar caps 89
Massena (New York) plant 57–60, 74, 111
Mayfly (dirigible) 71
McArthur, Warren 158
McKinley, William 77
Mellon, Andrew 3, 48, 51–52, 58, 80–81, 96, 103, 124–126, 152
Mellon, Richard B. 46, 48, 81, 96, 104, 111, 152
Mellon, Richard King 153–154
Mellon, Thomas 46
Mellon business model ("Mellon System") 48–49
Mellon family 46–48, 80–81, 145–146, 152–153, 174
Mellon Square 174
Mercedes-Benz 69, 73
Mercury 143
Merle Chemical 20–21, 35
Merrill, Neil 88
Metal spinning 83
Mies van der Rohe, Ludwig 87–88, 174–175
Minet, Adolphe 34
Mining 83
Mobile homes 204–205
Moissan, Henri 24
Moka Express coffee maker 166–167
Morgan, J. P. 2
Motor homes 176–180
Motorcycles 86
Muscle Shoals, Alabama 101–103
Musical instruments 88
Mussolini 166

Napoleon III (Louis-Napoleon Bonaparte) 11–12, 69
NASA 196–200

National Labor Relations Act (Wagner Act) 113–120
National Labor Relations Board 120–121
National Recovery Act of 1933 119–120
National Steel 2
New Kensington plant 46–49, 50–53, 57, 60–61, 67–68, 85, 89, 98, 106, 118–120, 153, 159
Niagara Falls (power plant) 15, 56–58, 60–61, 67, 111
Nitrates 102–103
North Carolina 73–75
Northern Aluminum Company 58, 65, 151–152

Oberlin College 15, 26–30, 76
Oersted, Hans Christian 10–11
Oligopoly (aluminum industry) 147–152, 187, 212–213
Ormet Aluminum Company 149

Packaging 6, 91, 223–224
Paris Exposition of 1855 12, 157
Paris Exposition of 1878 157
Paris Exposition of 1937 157
Park, James 38–45
Paternalism 94, 104–112, 188–189, 216
Pechiney, A. R. 74
Pechiney Aluminum 32–33, 35, 49, 74–75
Pelt, Mary 120
Pennsylvania Salt Company 61–62, 65
Perkin Award 63
Perkins, George 80
Permanent Employment Fund (of ALCOA) 111–112
Perry, Admiral 66
Pierce-Arrow 69, 87–88, 180–181
Pioneer (spacecraft) 2, 141–142
Pittsburgh East End 38–43, 51
Pittsburgh Reduction Company 42–49, 50–55, 58–62, 111, 117–120; founding of 42–43; growth 58–62; litigation of 50–55
Pittsburgh Testing Company 38–44
Pope Leo XIII 15
Power Canal 57–59
Pre-fab housing 172–174
Prices/costs 13, 65, 77, 125–126, 134, 138, 145, 165, 171–172, 188
Production (weight) 13, 44–45, 77, 94, 95, 134, 143, 148
Properties 5–8, 190, 205
Prouve, Jean 172
Pullman City 104–105, 110
Putin, Vladimir 212

Queen Victoria 15

Railroad applications/ cars 148, 169–170

Rapid transit 209
Recycling 194–196, 222–224
Red mud 61, 138
Research approaches 69–74, 76–79, 88–90, 126–127, 132, 154–157, 188–189
Revere Copper and Brass 149
Reynolds, Richard 91
Reynolds Metal Company 2, 91, 134–136, 144–145, 149–150, 175–179, 188–189, 193–196
Reynolds Wrap Aluminum Foil 150–151
Rio Tinto Canada 214
Rivets 141
Rogerstone (Wales) plant 151–152
Rolls-Royce 69
Roosevelt administration 127–130, 134–136, 144
RUSAL company 212–214
Russia 5, 167, 210–211
Russian aluminum industry 80, 86, 139, 167, 210–216
RV 179–180

Sagan, Carl 2, 141–142
Saguenay River 93
St. Lawrence River 57–49, 93
St. Louis plant 61–62, 64, 107–110, 122
Sales 83
Sample, William 42
Sand casting 72–74, 83, 89
Schmidt, Albert 33–34
Scott, Robert 42
Scottish industry 58–59
Shawinigan Falls (Quebec) plant 58–60, 64
Sheet 53–54, 66–67, 83, 91, 94
Shenandoah (airship) 72
Sherman Antitrust Act 64, 126–128
Shipbuilding 160–163
Sick Fund (of ALCOA) 111–112
Siding 177–179, 197–198
Siemens, August 21
Siemens, Charles 21–22
Siemens family 21
Silver 6–7
Skis 183–184
Skylab 196
Smallman plant 46–49
Smithfield Street Bridge 180
Social Security Act 114–115
Space applications 195–200
Speer, Albert 157
Sputnik 195
Stamping 53–54
Standard Oil 43, 127
Steel cans 195
Streetcars 169–170
Structural applications 180–181

SUAL (Siberian-Urals Aluminum Company) 213–214
Submarines 189–190
Subway cars 190–191
Sumitomo Metals 142
Surinam 138
Swanson TV Dinners 184

Taft, William 32–33, 65
Taft Decision of 1893, 54
Tariffs 77–78, 187–188, 210
Tennessee Valley Authority (TVA) 100–103
Tennis rackets 183–184
Tesla, Nikola 17, 32–34
Tesla Motors 218
Thaw, George 43
Thermic process 20–21
Thermochemical process 30–31
Tiffany's Department Store 15
Times Square ball 157–158
Times Square Building 163
Toronto Subway 190
Trailer parks 179, 203
Trailers 176–177, 204–205
Transparent Aluminum 216
Transportation 148–149, 160–162, 168–179, 221
Truman administration 144–146
Tubes 89–90, 161, 181
Tumblers 182–183
TV dinners 184–185
Two-piece cans 193–194

Unionization of aluminum industry 113–120
United States Aluminum Company 54
United States Steelworkers 114–120, 123
University of Michigan 2
University of Pittsburgh 39–40
Urals Aluminum Smelter (UAZ) 139

Verne, Jules 1, 8, 13
Vertical integration 59–61, 156–157
Vickers Company 71
Vietnam 225–226

Wallace, Donald 21
Washington Monument 13–16, 29
Wear-Ever Division 66, 83
Weimar Republic 79–80
Welding 207–208, 218
Wells, H. G. 13
Western Electric Company 27–28
Westinghouse, George 19–20, 27, 34, 40, 43–45, 104
Westinghouse Electric Company 19, 165–166
Wheels 201–203
Wilhelm II of Prussia 69
Wilm, Alfred 70–72

Wilmerding Pa. 105–107
Wilson, Irving 145–149, 152–154
Wilson, Woodrow 77–78
Wilson administration 101–103
Wilson T-2000 racket 183
Wilton Brass Company 198–199
Wires 205–206
Wisconsin Aluminum Fabricators 82–83
Wohler, Fredrick 10, 14, 28–29

World War I 67–73, 79, 101, 138
World War II 125–128, 134–136
Wright Brothers 52–53, 68

Yadkin River 74
Yeltsin, Boris 212

Zeppelin, Ferdinand, Count von 69, 161
Zeppelin airships 69–72, 161–165

www.ingramcontent.com/pod-product-compliance
Ingram Content Group UK Ltd.
Pitfield, Milton Keynes, MK11 3LW, UK
UKHW041938140426
5217IPUK00014B/545